"Solar ovens must surely run up there with the clothesline in making easy and free use of the sun."
— Stuart Ward

HEAVEN'S FLAME
A Guide to Solar Cookers

by Joseph Radabaugh

Home Power Publishing

Heaven's Flame
A Guide to Solar Cookers

By Joseph Radabaugh

Second Edition, Second Printing

Copyright © 1998 Joseph Radabaugh, all rights reserved. No part of this book may be reproduced or utilized in any form or by any means, electronic or mechanical, without express written permission from the publisher, except in the case of brief quotations embodied in critical articles for the purpose of review.

No liability is assumed with respect to the use of the information herein.

All photographs have been used with permission and are copyrighted in the name of the original photographer.

Cover art and illustrations by Benjamin Root.

Radabaugh, Joseph
   Heaven's flame : a guide to solar cookers / by Joseph Radabaugh ; [art & illustrations by Benjamin Root]. — 2nd ed.

   p. cm.
   Includes bibliographical references.
   Preassigned LCCN: 98-
   ISBN: 0-9629588-2-4

   1. Solar ovens—Design and construction. 2. Solar cookery
   I. Root, Benjamin. II. Title.

TS425 R331998 683'.8
             QBI98-804

Direct all inquiries to:

Home Power Publishing
PO Box 275, Ashland OR 97520
Website: www.homepower.com

Printed in the U.S.A.

Dedication

I'd like to dedicate this book to my children, Promise, Toketee, and Abraham, to their grandchildren, and to all the children of the earth.

To those who are willing to make a personal effort to create a new future, where hope still blossoms.

To those who have dreams and visions of a healthy planet and a sane, compassionate human community.

Acknowledgements

I'd like to thank all the people who freely submitted material about their work with solar cookers. This book could not have been possible without them. With their help we are able to see our work put together to make a complete statement. I hope this book succeeds in this important task.

Table of Contents

Chapter 1:   A History of Solar Cooking............. 1

Chapter 2:   The Environmental & Social Case
for Solar Cookers..................... 5

Chapter 3:   Adventures in Designing.............. 11

Chapter 4:   Questions & Answers 21

Chapter 5:   The Work of Others:
Designers & Promoters 35

Chapter 6:   Designing & Building
Your Own Solar Cooker............... 71

Chapter 7:   Building the SunStar................. 87

Chapter 8:   Using the SunStar 111

Chapter 9:   Cooking Tips for Various Foods 117

Chapter 10: How to Produce the SunStar
in Kit Form 127

Chapter 11: The Rainbow Cooker:............... 131

Access 137

Index 141

Preface

These are exciting times for those involved with solar cookers. Until recently, most of us were following our own ideas and goals without knowing what other people were doing. This was good since it allowed us a sense of personal freedom to define our individual visions. This creative freedom shows up in the diversity of designs now ready for the attention of our human community. Now, those of us involved with solar cooking are reaching a critical stage in which we are experiencing the effect of synergy. Each person's success creates a ripple effect being positively felt by all others involved. For the first time, we can see our individual works coalescing to present a complete picture of the potential of this simple solar tool.

I have a few goals in writing this book. One is to tell you that solar cookers work. Another is to share with you the variety of cookers available. I also want to give a few examples of how people are sharing these tools with others, bringing you up to date with this relatively unknown movement that is beginning to experience rapid growth. This is a true grassroots movement that needs you and your special skills. I'll attempt to speak for all involved with solar cookers and extend an invitation to you to join us in any way you can.

Another goal is much more personal. While playing with solar cookers as a hobby, I came up with what I believe is an important addition to the collection of solar cookers. My design, the SunStar, is a very hot and efficient cooker made out of the barest of materials. The design is ready for serious evaluation by a world audience, including you and your friends.

In this book I will share with you all I've learned about building and using the SunStar. Long ago I made the decision not to patent this cooker. Instead, through the publication of this book, the SunStar is in the public domain. This means it is yours. If you find a way to get the design out, you have not only my blessings, but my sincere thanks. Our brothers and sisters throughout the world need these cookers now, as a survival tool. All the people of the world deserve to know that simple solar cookers exist.

The first edition of this book had some intangible goals written between the lines. I wanted to inspire people to follow their own vision and dreams, to work on them, refine them, and take the time to share them with others. I often told people that I saw solar cookers as an oboe in an orchestra. By itself, it makes a rather odd sound. But in tune with the other instruments, it can lift us to new inspired heights. I hope that solar cookers are coming in tune with a whole host of other positive visions.

In light of the times in which we find ourselves, this edition is a little more blunt about the thoughts that inspired this book. We are many on this earth, and many more are coming, which is an overriding concern of the day. Our sheer numbers are causing many new social and environmental strains. Solar cooking addresses real problems with a real solution.

– Joe

Chapter 1

A History of Solar Cooking

For most of human existence, the cooking of food was unknown. People ate food in the condition in which they found it. Then humans found that fire could be controlled and used to cook food. Fire is essentially solar power stored in the form of wood. If we look at it this way, solar was the first method of cooking on earth.

Looking for the beginnings of what we now call solar cooking, we find some isolated stories in the distant past. The Essenes, an early sect of Jews, gently heated wafers of ground sprouted grains on rocks heated by the desert sun. This was not cooking in the contemporary sense. The goal was not to overheat the wafers, but to heat them only to the point which did not kill the living enzymes in the grains. This created a food source that is extremely healthy for the human body.

The first known person to build a box to solar cook food was Horase de Saussure, a Swiss naturalist. He cooked fruits in a primitive solar box cooker that reached temperatures of 190°F. He was the grandfather of solar cooking.

During this time, others also started using solar cookers. In India, a British soldier patented a fairly sophisticated solar cooker that looked a lot like the Solar Chef, which is described later in this book. In 1894, there was a restaurant in China that served solar cooked food. There are also stories about an early sea captain who created a solar cooker he could use on long voyages.

Cookers we see today started evolving in the 1950's. Our world was still reeling from the insanity of war. People were looking for ways to create a stable and peaceful future. The worldwide nature of the previous war showed, in some ways for the first time, that we are a world community facing global problems that affect us all.

One of these global problems was the growth of deserts around arid communities. The United Nations and other major funding agencies initiated many studies to design solar cookers that could alleviate some of the reliance on plant life for fuel. Many top engineers of the 1950's were hired to study different aspects of solar cooking designs. These studies concluded that properly constructed solar cookers not only cooked food thoroughly and nutritiously, but were quite easy to make and use.

The UN then sponsored studies and programs to introduce these cookers into cultures where the need was most apparent. These efforts proved mostly unsuccessful. In one study, 500 wooden solar cookers were given to a refugee camp. Three months later they had been chopped up and used for firewood. The social scientist concluded that traditional cooking methods were too culturally ingrained, and people were unwilling to adapt.

The UN did note one success. In a northern Mexican community lacking fuelwood, they found that the cookers were still in use five years later. This showed that it was possible to get cookers out to people in need.

The UN, in spite of this one success, concluded that solar cookers were not a viable option and all funding for solar cookers stopped.

Many involved with this early effort concluded that the studies themselves were flawed. They felt the designs being promoted were too complicated. Also, the cookers were too expensive for the intended users. They felt that more work was needed on the cooker designs. A few of them kept the potential of solar cookers alive by continuing to develop them in their own sunny backyards.

Others felt that the promotion techniques used in the UN studies were also flawed. Social scientists, who had never integrated solar cooking into their own lives, were in charge of the UN studies. The cookers were promoted as a

Chapter 1: A History of Solar Cookers

solution to poor people's problems, but certainly not as cooking tools that would be useful in developed countries. This caused solar cookers to be looked upon as a second class tool by those being asked to use them. Solar cooks sought new ways to promote solar cookers that were more sensitive to the cultures they were trying to share them with.

This is the history we are now writing. Gone is much of the heavy engineering in favor of backyard hobbiests building their own simple cookers. Gone are the paternalistic promoters telling people what they need to use. Instead, people are sharing a tool that they themselves enjoy using. This is a great maturation within the solar cooking movement. It is also the reason we are now on the verge of a new growth cycle. The future story of solar cooking is wide open.

Heaven's Flame

Chapter 2

The Environmental & Social Case for Solar Cookers

It is often people's concern for the environment that compels them to start using solar cookers. In a world full of energy problems, it seems so obvious that using the sun's free rays to cook food will help. Because learning to use solar cookers takes effort, it is best if we understand why the use of solar cookers is important. Reasons include:

1. Decreasing firewood and fossil fuel consumption. This could result in decreased pressure on fragile ecosystems and economic systems of many human communities. Economic problems often lead directly to environmental problems.

2. In some areas, solar cookers will decrease the use of dung as a cooking fuel, which could then become available as a fertilizer.

3. Carbon dioxide release is driving the disruption of our global weather patterns. Solar power creates no carbon dioxide release. Also, every plant not burned can consume carbon dioxide.

4. Burning fossil fuels releases a wide range of pollutants, adversely affecting the health of individuals. Solar cooking does not burn these fuels.

5. The use of solar cookers reduces time and effort spent to find or buy fuel. This can measurably affect the quality of life.

6. The cookers can purify biologically polluted water by pasteurization. Biologically polluted water causes many diseases and human misery, killing millions annually.

Heaven's Flame

In 1992, the United Nations gathered leaders of the world together in Brazil for the Summit on the Environment. The Studies presented predicted that cooking fuel shortages were going to be one of the main problems facing our world in the next century. They estimated that 2.5 billion people will experience severe shortages by the year 2000. Also, nearly 50% of the rainforests being cut down are used just for cooking-fuel.

As the world leaders discussed these problems, not once did they mention solar cooking as a solution. This is a glaring oversight, especially at a summit sponsored by the United Nations which conducted the first extensive studies of solar cooking. If they had looked at their own earlier conclusion, that solar cookers work and are easy to build and use, these leaders would have realized that solar cookers deserve, and that our times demand, a seriously renewed effort. Instead, our leaders are missing one of the potential solutions to the shortages of cooking-fuels facing over two billion people.

At the summit, leaders crafted important press releases about their concern for the world's environment, while at the fairgrounds a few miles away, members of Solar Cookers International were cooking and giving away solar meals. This obvious connection is still not being made.

A few stories illustrate the real need for solar cookers. At the first World Conference on Solar Cooking, I was listening to a woman relating her efforts in sharing solar cookers in Pakistan. She said that the day before she arrived at one of the villages, a mother had committed suicide because she could no longer find wood to cook her family's meals. This suicide took place in the village square as an act of protest—but also one of ultimate desperation. How many more families are experiencing this kind of desperation?

People working with the Afghan refugees in Pakistan found it was costing more to cook the food then the cost of the food itself. Relief agencies estimated that for a family of eight, it took 99 hours to collect the firewood to cook food for one week. This is a real and growing problem. Only four percent of Pakistan is still forested, and these areas are disappearing at an alarming rate.

Chapter 2: The Case for Solar Cookers

In Nepal, very few forests remain. Yet, for rural families, wood for cooking still represents 98% of their energy demands. The loss of healthy forest ecosystems in the Himalayas increases the severity of the annual floods and destruction downstream in Bangladesh. The ecosystems of the Himalayas are not going to stabilize until we find some way to remove these pressures. This Himalayan country is showing signs of a potentially horrible ecological collapse.

At the same time, higher up on the Tibetan plateau, a great solar cooking success story is taking place. Solar cookers are commonly used cooking tools in Northwestern China and Tibet, areas of very little wood and an abundance of sunshine. The Chinese government chose to offer solar cookers at reduced prices to people in these areas. By subsidizing cookers, Tibet is becoming one of the first solar cooking cultures on earth. The tools were made available, and the people are ready to use them.

Another example of the need for solar cookers comes from the western hemisphere. This story is about a village in Haiti, which for generations had lived in balance with their ecosystem. They had a beautiful rainforest hillside behind them, which supplied wood and wild plant foods. A river fed their lagoon that teemed with aquatic life, supplying the village with an abundance of food.

Left: Rudecinda and her children in El Salvador. Solar Energy International taught solar oven building in this rural repatriated refugee community.

Above: In Kenya, where wood for cooking is scarce, women in Dadaab refugee camp plant trees to earn solar cookers. Solar Cookers International trains women to use the cookers, who then train their neighbors.

Then, during recent political and economic upheavals, the cities found their normal supplies of cooking fuels cut off. They started demanding charcoal from the countryside. This resulted in the over-cutting of the forest above the village. When the rains came, silt and mud flowed into the river and lagoon. The siltation of the lagoon essentially wiped out the aquatic life that had sustained the village. Faced with no way to survive, the people in the village pooled their money to build a boat to send their strongest children to Florida. This was one more act of ultimate desperation caused by the need to cook food.

Look at the Sahara desert. It continues to expand due to pressure on the plant life. There are areas in the northern Sahara where the people can cook only a couple of hot meals a week due to the lack of firewood. This is in an area of abundant sunshine. Families are walking up to 20 kilometers every day for wood. In their desperation, the villagers are even cutting living trees and bushes. These are seeds for total environmental collapse. If the world com-

Chapter 2: The Case for Solar Cookers

munity does not seriously address this problem, this desert will continue to expand, dislocating many surrounding population centers. The people of the whole continent of Africa need and deserve these tools. One billion pounds of wood are used each day for cooking in Africa. An area the size of Wisconsin is burned every year. The effort it takes to collect this wood is an enormous drain on both the ecosystem and the communities living there.

People who have been to Africa to share cookers are coming back and telling of the many hardships involved with the gathering of firewood. In areas of war or unrest there is a constant danger of being robbed, shot, raped, or kidnapped, being faced by individuals who have to roam far from camp for wood.

Refugee camps are often placed in the territories of other people who do not want the refugees collecting firewood from the local environment. So families must trade some of their food allotments to the neighboring villages for wood. There is a real need for solar cooking devices in these areas.

Using solar cookers to pasteurize water is relatively new and of obvious worldwide importance. The drinking of biologically polluted water in developing countries results in sicknesses that are the leading cause of death among young children.

Direct environmental effects are felt in areas that use firewood to boil drinking water. It takes approximately 1 kilogram (2.2 pounds) of wood to boil just one liter of water. This significantly contributes to deforestation, pollution, and other forms of environmental degradation.

There are also direct monetary costs related to poor drinking water. In the recent outbreak of cholera in Peru, the Ministry of Health urged all residents to boil their water for 10 minutes. The cost of doing this amounts to 29% of the average poor family's income. In Bangladesh, boiling the water takes 11% of the income of a family in the poorest quarter of the population. In Jakarta, Indonesia, more than 50 million dollars per year is spent on boiling water. In Cebu, Philippines, about one half of the city of 900,000 boil the water that they obtain from their unreliable chlorinated water system. These costs are direct drains on these societies. The use of solar power for pasteurization could partially alleviate these costs.

At The World Conference on Solar Cooking, Shyam S. Nandwani of Costa Rica presented figures on some of the savings we could expect from the use of solar cookers. The numbers are based on families of four to five people using the cookers for an average of seven months. Annually, a family would save 210 liters of propane, 648 kilograms of firewood, or 1,160 kwh of electricity. Where I come from, in California, electricity is 8 cents per kwh, so the savings would be almost 90 dollars each year, enough to build a few good solar cookers.

If only 1% of the 1.5 billion people affected by cooking fuel shortages today were to use solar cookers seven months a year, they would save two million tons of wood. This would also prevent the release of 85,000 tons of pollutants, such as SO_4, CO, NO, ash, etc. These savings represent the equivalent of 10 million trees a year.

There is also an environmental case for people in developed nations to use solar cookers. We may think our cooking fuels are cheap and plentiful, but maybe we should take a second look at some of the environmental impacts of the fuels we use. Oil and gas have their spills and pollution. Nuclear power has its waste and accident problems. Hydro power disrupts ecosystems. These costs are by no means cheap. "Plentiful" is a questionable term for finite resources. The sources of cooking fuels you use now, if not solar, are ecologically and economically unstable. Their prices will escalate. Solar cooking lets you take some control of your personal energy needs, using abundant free energy on sunny days. It is time for those of us in the developed world to take responsibility for living on earth. I wish you could realize the power we in the developed nation have. We have the money and resources to help start this solar industry.

To change the situations in many other countries, we need to offer people good solar cookers at a price they can afford. As you will see in this book, effective solar cooking tools are easily available.

Chapter 3

Adventures in Designing

Many people working with solar cookers have loosely followed a similar pattern which led to their growing involvement. There was no cultural experience of solar cooking from which any of us could draw. I'd venture a guess that none of us were raised on solar-cooked foods. At some point we heard or read about solar cookers. Maybe we met someone using one. At first, most of us had a hard time grasping the significance of this natural phenomenon. More likely we felt a simple curiosity at seeing how easily the sun could be harnessed to do meaningful work.

Each of us caught our own personal glimpse of the potential of solar cookers. We began to imagine a world where solar cooking was common. Thoughts and visions, coupled with a growing familiarity of solar cooker use in our own lives, naturally led many of us to our expanding involvement with solar cooking.

This section shows how my own involvement in designing ovens grew, and eventually led to the SunStar solar cooker. Hopefully, it will give you a sense of the work that has gone into each of the many good solar cookers being presented in this book. Also, if you wish to design your own solar cooker, this story might unveil a pattern of problem-solving which you may experience.

The Friend

As a friend was getting ready to enter the Peace Corps, he dropped a solar cooker off at a cooperative where I lived. He had built it from plans he found in *Mother Earth News*. I wished him well and set the cooker in the backyard.

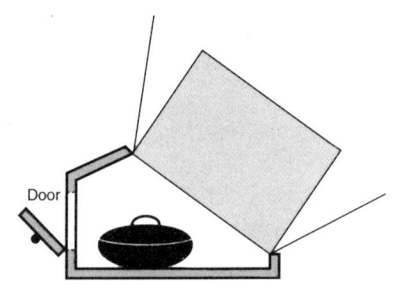

Above: The "Friend's" oven.

Before the weeds grew up to hide the solar cooker completely, another friend decided to try it out. He proceeded to bake some pretty funky bread, but also some remarkably good potatoes.

David was studying to become a mechanical engineer, so the angles and the simple use of solar energy caught his imagination. His growing interest in this solar tool started to pull on me. I was the gardener of our group, so I began solar cooking the vegetables from our garden. While using this cooker in the garden, I caught a first glimpse of the potential of solar cookers.

This solar cooker was built out of thin plywood and insulated with one inch of fiberglass. The cooker used aluminum sheets recycled from newspaper printing presses, and painted black, for the inside walls. There was a removable door placed at the back of the ovenbox. This cooker used four square reflectors to direct sunlight into the ovenbox. A confusing series of sticks and strings stabilized these reflectors. The reflectors also folded down onto the glass for storage.

This cooker worked but only reached about 225° F (107° C). Because of obvious design problems, this cooker stimulated a natural curiosity to find out what a solar cooker could do when designed properly.

One of the problems with this cooker was the glass, which kept cracking in one corner. After replacing the cracked glass a couple of times I noticed that the frame holding it was binding in that corner. This seemed to be putting an unequal tension on the glass. After loosening the corner, the glass didn't crack again.

Why glass sometimes cracks has been a subject of many of my conversations with solar cooking people. Most agree that cracking is due to temperature related stresses. In the case of this cooker, unbinding one corner relieved the unequal tension and solved the breakage problem.

Chapter 3: Adventures in Designing

The window of this cooker also steamed up. The steam buildup on the window interfered with light penetration which led to lower cooking temperatures. I started using pots with tighter fitting lids and this helped somewhat. Then I added a second pane of glass with a small air space between the two, creating a thermal pane. This solved the steam problem because the surface temperature of the inside glass rose high enough so water wouldn't condense on it.

The door on this cooker was too small for convenient use. The reflector system was tedious to set up, and near impossible to stabilize in the wind. The insulation did not seem thick enough, because one could feel hot spots on the outside of the ovenbox. With these initial experiences, I decided to design my own.

The Greenhouse Oven

The first cookers I built were a series of prototypes, large and small. They looked a lot like a solar greenhouse, so I started calling them Greenhouse Cookers.

A variety of materials were used for the oven walls and reflector backing: cardboard, paneling, thin plywood, aluminum sheeting, and sheet metal. These cookers used fiberglass household insulation which was abundant around my workshop. I increased the insulation thickness to 2 1/2" (6.5 cm.), hoping to increase the heat retention of the ovenbox.

Most of these cookers had thermal pane windows. I liked the extra power that two panes of glass gave to the oven.

I experimented with the placement of the door, sometimes putting it in back of the oven and sometimes using the window as a door. The back door made sealing the glass on

Below: The "Greenhouse" oven.

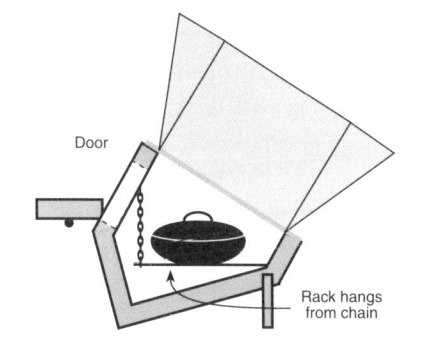

the oven box easier. Using the back door also allowed me access without moving the reflectors. They were secured in place with bolts which took time to remove. Also, I was using a highly reflective aluminized plastic film for the reflectors, so the glare hurt my eyes when I stood in front of the oven. The back door access, when built large enough to be convenient, became my favored choice. (You will see that this choice did not last. My own cooker, the SunStar, uses the window as access, because the problems listed above were eliminated.)

These ovens tilted to focus on the sun in its various positions. Tilting created two design problems. One was a rack to keep the food level. The other was a stand to hold the oven in various tracking positions.

After a few tries, I ended up using a rack connected to the ovenbox near the bottom of the window. A chain and hook held the other side of the rack in the different positions. This rack worked, but was inconvenient. When adjusting the oven to a new sun angle, the food rack needed adjusting at the same time which was difficult to do with hot food.

A few stands were developed to tilt the ovens. In some cookers, adjustable legs propped up the oven in different positions. In others, a stand cradled the cooker in different positions.

These cookers used four equal-sized reflectors that folded down on the glass when not in use. Usually, triangular reflective pieces were fitted between the four main reflectors, increasing the amount of light entering the ovenbox. This system was stabilized using bent metal brackets, nuts, and bolts. This method was not satisfactory, because it was too labor intensive to repeatedly tighten and loosen bolts.

These ovens cooked food quite well. The temperatures reached over 350° F and the thick insulation kept the outside of the box cool. By tilting the oven I was able to continue cooking until sunset.

These designs were very good first steps. Still, the solar cooker that I wanted to design needed to be easier to build and more convenient to use. Also, even though no one part was expensive, collectively the cost of my cooker

Chapter 3: Adventures in Designing

designs were not cheap. This was also true with the weight of the cooker. The parts kept adding up until these cookers were somewhat heavy. Back to the drawing board.

With a design inspiration, I created a solar cooker in which the stand and the rack were fixed permanently together. The ovenbox and reflectors pivoted around the rack. This made it easy to focus on the sun without spilling the food. I had something unique here and proceeded to build a few prototypes which worked very well.

At that time, I was studying soil science at a California university. In the library, information on solar science happened to be next to my soils information, so I began to see what type of research was available on solar cookers. Surprisingly, there was quite a bit.

The work of Maria Telkes stood out among these papers. Her favorite cooker looked remarkably similar to the design I was building. She designed her ovenbox to pivot around the rack too.

Through her writings and research, Maria Telkes became my mentor. She was a top solar engineer during the 50's and 60's. Her research was funded by major grant foundations. She worked closely with the United Nations as they studied the potential of cooking with sunshine.

Maria Telkes studied the effects of many design variables. For example, she studied the results of using one, two, three, and even four panes of glass on the oven box. She concluded that two panes created the maximum power potential for solar cookers. She also found that an oven box which pivoted around the rack answered most of her design requirements. This was to become her signature cooker.

Her main goal in designing solar cookers was to make them affordable to the very poor. To this end, she built a series of ovens using woven baskets lined with straw and a thin coat of mud.

Maria was an engineer, not a businesswoman. Much of her solar oven research was geared towards studies, not potential commercialization. Still, she concluded that even very efficient ovens can be inexpensive if mass-produced.

Heat Storage

Maria Telkes' involvement with solar power extended far beyond solar cookers. One area where she did a lot of work was with methods of storing heat. Solar power might be abundant in the day, but when the sun goes down, stored power would be useful. She studied many potential heat sinks. Heat sinks are materials that can store heat for future use. She developed one unique solid heat sink, using a mixture of the common chemical compounds sodium sulfate and calcium sulfate. It is a simple material, capable of storing significant quantities of solar energy.

What made this material unique was that it went through a solid to solid phase change, and stored up large quantities of heat. Most substances go from a solid to a liquid, or a liquid to a gas during the phase change. For example, liquid water rises in temperature to the boiling point and then sits at this temperature absorbing large quantities of energy until it makes the phase change to a gas. The material Maria found went from a solid 5 sided crystal to a solid 6 sided crystal. This makes it a very useful heat sink because it doesn't need a leak proof container.

Also, the temperature at which this phase change takes place can be regulated from 350° F to 500° F, depending upon the mixture of the two chemicals. Though this heat sink has limited use in small solar ovens it could have great potential for community solar bakeries. The heat sink could be placed in a large cooker and with added solar power the heat sink will rise in temperature towards the phase change. It will sit absorbing energy at this temperature. Once it has stored up enough energy to complete the phase change to the different crystalline structure it will continue to rise in temperature. As this heat sink cools, it will lower to the temperature of the phase change where it will then stay, giving off energy at a constant temperature, perfect for baking. In a solar bakery, this substance could be used to maintain a constant temperature in the baking range. The formula for this heat sink can be found under patent number 2808494 (Apparatus for Storing and Releasing Heat).

Chapter 3: Adventures in Designing

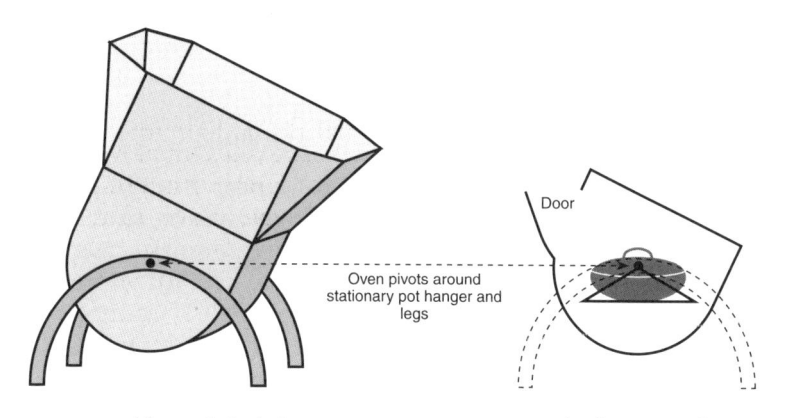

Above: Maria's oven uses a common axis for oven tilt and pot rack leveling.

During this research period, I found that most university libraries held a wealth of information on simple, alternative tools. This is especially so if you look into the studies from the 50's to the early 60's. One good study is the Proceedings of the United Nations Conference on New Sources of Energy, Rome, 1961. Back issues of *Solar Energy* magazine from this period are also great information sources. There is information in our libraries on many other simple tools which, like the solar cooker, deserves renewed attention.

Fortified with this new research, I started designing and building ovens again. Most of my ovens had used four equal sized reflectors with four triangular pieces connecting the sides. Because of stability problems, I never felt comfortable with this style of reflector. Pondering this problem, another design inspiration hit. Four trapezoidal reflectors, by virtue of their shape, are much more stable. (I describe this later under "Reflectors" in chapter 6).

With this new reflector system and Maria Telkes' pivoting design, I felt that I had an oven that could be commercially successful. I built a couple dozen of these solar cookers out of sheet metal and wood. This is also when I learned of the frustration in trying to market solar cookers to a culture not aware of their potential. They were visually appealing and worked well. Because of the material and labor costs, though, these solar cookers needed to be priced at around $150. I tried to sell the ovens, but most people were not willing to pay that kind of money for something they weren't sure that they would use. A few of these cookers were sold, and the rest were given to friends.

By this time, I was firmly committed to promoting solar cooking. I just needed a design that wouldn't threaten people's pocketbooks.

Missing Link

My new design goal was to make ovens that would cook for a small family and cost under $30. Actually, my goal was to let the cooker cost no more than an average bag of groceries. This price would allow most people to experiment with solar cooking. It would also demand a quantum leap in simplification because most of my previous ovens had more than $30 in material costs alone.

They also needed to be very light weight and portable. When cooking with a solar cooker you may have to move it around to avoid shadows. Moving an oven that weighs 20 pounds may sound easy, but if it's filled with pots and boiling food it becomes quite heavy and awkward.

I was looking for some kind of missing link, an oven that beginners could use successfully.

During the summer of 1987, I built two ovens, identical except for the insulation. For comparison, one was insulated with cardboard layers. The other used fiberglass. If only a small part of the power of the cooker was lost by substituting cardboard for regular insulation it might be worth it, due to the cost, simplicity, and ease of using cardboard. I built these ovens using simple materials, cardboard and aluminum foil. Not expecting the cardboard to work, I built these cookers with thermal pane glass.

I drove up to a mountain lake and set out the cardboard insulated oven first. While putting the final touches on the fiberglass model I went over and checked the thermometer on the cardboard solar oven. It was approaching 400° F after only 20 minutes of bright mountain sunlight. I was awestruck. This was my missing link. I never did finish setting up the other solar oven.

Chapter 3: Adventures in Designing

The oven temperature, with continued focusing, soon reached 450° F and the box began smoking, (the flash point of paper is 451° F [232.7° C]).

Instead of letting the oven burn, the thermal pane was replaced with a single piece of glass and tried again, refocusing the still empty box in the same bright sun. Now the ovenbox reached 390° F. It was one of the hottest ovens I had ever made and yet well within the safety margin. Note that continually focusing a solar oven without food inside to see how high the temperature can rise is not a normal way to use a solar cooker. With food inside, the temperature will stay much lower since the food absorbs much of the heat keeping the temperature nearer 212° F (100° C).

With some scrounging, I was now able to build a family-size oven for less than five dollars. This oven, built with recycled and recyclable cardboard boxes, glass, and aluminum foil, is light weight and portable. Named the SunStar, this cooker will be described in detail in Chapter Seven.

Left: The missing link, a.k.a. the "SunStar."

Heaven's Flame

Chapter 4

Questions & Answers

Once I had developed the SunStar solar cooker, I needed to find a way to tell people about it. I drew up some simple plans to distribute. In the summer of 1987 there was a Harmonic Convergence gathering on Mount Shasta. I took my new cooker up to Panther Meadows and proceeded to cook meals and give away plans.

After this, I decided to go south to Redding and set a cooker up at the Jolly Green Giant swap meet. After a day of cooking and giving out plans for a dollar (or free) in 110° weather, I ended up with a slight case of heat stroke and 19 dollars.

Sharing in this way was a lot of fun though, so I decided to refine my plans and presentation and go on a swap meet tour of the southwestern United States. Later, I started showing the oven at alternative energy shows, craft fairs, farmers' markets, barter fairs, and Rainbow Gatherings.

These gathering places were very useful for promoting solar cooking. I could reach many people who normally would not be exposed to solar cooking. Another nice thing about these shows is that they are world wide. Almost universally, there are farmers' markets that are prime places to educate people about the value of solar cooking. I have talked to thousands of people in these settings.

There are some simple guidelines in sharing cookers in these situations. One is to only share these cookers on sunny days. I found that setting a cooker out on a cloudy day invited too much ridicule. Over the years, as my confidence has grown in the use of these tools, I've started to bend this rule. My confidence, with some effort, can overcome this natural disbelief syndrome.

Another guideline is to show these tools only when people could see something actively baking or boiling. Early in the morning I would set up a cooker with dried beans and water into one of my cooking jars. I'd keep the solar cooker focused on the rising sun until I could start to see small bubbles rising, indicating the first stages of boiling. Then I'd start my solar show.

When showing the cookers, I keep the beans boiling as fast as possible for the demonstration effect. Some of the water necessary to cook the beans will boil out, so I add more water. I put another jar with some more water in it next to the boiling beans. Then I can pour preheated water into the beans, allowing them to continue boiling. This is not the way I normally cook beans. It is a method I use for shows when I want the beans to boil actively for a long period of time.

I sometimes add more cooking jars with grains and veggies, so people can see the cookers cooking full meals. In another cooker I would bake breads, muffins, cinnamon rolls, or cookies, to show that solar cookers can bake.

It is good to offer people some information on how to build and use solar cookers, something that they can take home with them. For my own cooker, I refined a one page plan to offer for a dollar (or free).

These guidelines may be useful to you, because we are in an educational phase of solar cooking. Learning how to share this information effectively is important.

From talking to people at these shows, I compiled a list of the most commonly asked questions. With these questions and their answers, I hope to give you a little better understanding of the theories behind solar cooking.

What is it?

This question invariably comes as I set up the oven. "Why, it's a solar cooker," I'd answer, and then watch as bewilderment set in.

Chapter 4: Questions & Answers

Most people had never heard of solar cooking; it was a totally foreign concept. Maybe one in ten had heard of them, a few had seen them before, and every so often a rare passerby had used one.

Does it work?

This was usually the next question fired at me. With my children next to me, whom I'd raised on thousands of solar cooked meals, I'd say, "Yes, they work, and work very well."

What can I cook in a solar oven?

Solar cookers bake, boil, and steam foods. You can cook almost anything: whole grains, beans, cakes, breads, quick breads, pies, cookies, veggies, pizzas, stews, casseroles, and on and on. Solar cookers can also lightly fry foods. If you need to actively fry, you will need a solar parabolic cooker as discussed in the next chapter.

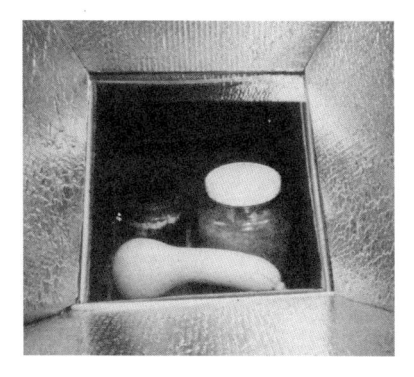

Above: Rice and veggies are among the many possibilities.

Does the food cooked in the sun taste different?

All people who regularly use solar cookers unanimously answer yes to this question. Once you learn how to use them, it will be the subtle flavors that will help convince you. It is the slow, even rise in temperatures that give the complex carbohydrates time to break down into simple sugars that account for this difference. A friend of mine is a gourmet cook at a fancy restaurant, and he concurs. He brings food slowly up to cooking temperatures to get the most flavor.

Others believe it has something to do with the prana or life force associated with cooking in the sun's rays. Whatever the reason, once you become a solar cook you will notice some of the best flavors you have ever tasted.

How long does it take?

This is one of the most often asked question and the one I have the hardest time answering. There are many variables that effect cooking times, including the differences between cooker designs and sizes, the quality of the sunlight on a particular day, the types and quantities of food being cooked, how often the cooker is refocused, and so on.

People then would often ask, "What about those beans you are cooking, on a day like this?" My answer would usually mention that I don't wear a watch, and anyway, beans taste better the longer you cook them.

Pressed a little further, I'd point to the sun. "If I started them here they will be done when the sun reaches there," as I pointed to another spot on the sun's path. I'm not trying to be evasive here, but truthful. Timing comes from experience gathered from your own cooker in your own area. It's best if you get one and learn timing on your own.

Most people don't like this answer. Then I say it takes just a bit longer than regular cooking methods. They are usually satisfied with this answer.

How hot does it get?

The majority of traditional solar cookers will reach temperatures between 250–400° F (130–200° C) with no food inside the ovenbox. When food is placed in the oven, the temperature will decrease at first, then start to rise slowly. When this rise takes place, it shows the food is nearing the cooking point. The temperature will not reach the same high as in an empty box. All food contains some amount of water, and water won't go above the boiling point. It is released as steam which carries away the excess heat.

You'll learn to use recipe temperatures as timing guides. It's best if you simply put the food in your oven and see how long it takes. Thermometers in solar cookers are not important, experience is.

How much food can I cook?

Most designs in this book cook plenty for a small family.

Chapter 4: Questions & Answers

Above: Tom Burns' huge "Villager" oven next to a more traditionally sized "Sun Oven."

Multiple ovens, or larger ovens, can be built for larger families or groups. The quantity of food you can cook will depend on many variables. There are environmental variables that affect the power of the sun on any given day. There are differences due to types of food. Large amounts of quick-cooking food can cook in the same time as small amounts of slow-cooking food. The design of your oven will also affect the amount you can cook. Solar cookers that intercept more sunlight will cook more food. Also, you will be limited by the cooking area. Some cookers hold more than one pot. Some do not.

Do they only work in summer?

No, but there are more cooking hours available in the summer. In the late spring and early summer, cooking hours might stretch between 8:00 AM and 6:00 PM. During the fall and early winter, effective cooking hours might only be from 10:00 AM to 2:00 PM. Nevertheless, there is something very special about cooking a meal on a bright winter solstice day, welcoming the return of the sun. I've cooked some excellent split pea soup in the middle of winter with snow on the ground.

Does it have to be hot outside?

The outside temperature does affect the solar cooker, but not as much as you'd think. The ovenbox is insulated from the outside temperature. When the oven is focused on the sun, its inside temperature will eventually reach a balance with the outside air temperature. Then it will gain a degree inside the cooking area with each degree gain of the outside temperature. Let's say that one oven reaches 330° F on a 60° F day. That same cooker will reach approximately 360° F on a 90° F day. Both of these temperatures will cook food just fine.

Above: An oven cooking in the snow at 11,500 feet in Colorado.

The most important factor in solar cooking is the brightness of the day, not its heat. Many times, a 50° F clear, low humidity day will cook food faster than a 100° F high humidity day. This is because of the difference in the quality of the sunlight reaching the ovenbox.

Should I preheat the oven?

This will only be necessary if you have a thermal mass or heat sink (rock, dark brick, iron bar, etc.) inside the oven to store up the sun's power, or if you made the inside of your ovenbox out of relatively heavy materials.

If you preheat the oven without thermal mass inside, the air in the oven may rise to 300–400° F (150–204 degrees C). Air, though, has very little ability to store heat. When you open the door or lift the glass to put food in, much of this heat will escape. So, without a heatsink, just put food in the oven and set it in the sun. No preheating is necessary.

Chapter 4: Questions & Answers

Can I bring the food to a boil first and then put it in the oven?
Yes. This can be a good way to shorten the cooking time, if desired. However, you will lose some of the unique taste of foods developed by the solar cooker's slow rise in temperature. Because of this, I rarely preheat foods.

Do I have to stir the food?
Another virtue of solar ovens is that it's nearly impossible to burn food. No stirring is necessary because all the food rises in temperature together so there are no hot spots that need stirring.

What is the ratio of grains and beans to water for solar cooking?
The best advice is to start with the normal ratio. This is one area in which it will be important for you to learn quickly with your own oven, and to adjust the ratio as your own experience guides you. If you don't have enough water, the food will dry out on top and not cook correctly. If your food runs low on water, you will have to open the box and add more water to the hot pot. Putting in a little extra water will slightly lengthen the cooking time, but the food will still cook.

Most promoters of solar cookers advise a little less water than normal ratios. I'm not comfortable with this answer. I feel that it depends on experience with your own oven. Some ovens use less water, and some use more. This is due to design differences, as well as differing cooking methods. For example, if I put beans and water in the cooker and it rises to a rolling boil, it will lose quite a bit of the water to steam. If I add more food after the beans begin to boil, or let the cooker go slightly out of focus, the beans would ease to a simmer and lose less water to the air.

If I don't eat my main meal in the daytime, can I enjoy hot meals in the evening?
Yes, you can easily cook meals in the early afternoon and store the food in the insulated ovenbox. Cover the glass with something that insulates, like a towel. If you are using

the SunStar, the cardboard reflectors fold down on top of the glass. Using this method, I have eaten hot solar cooked meals late in the evening.

Often, when I am unsure of the exact dinner menu for the evening, I just cook rice and/or beans in the solar cooker. If I don't eat the food that night, I put it in the fridge. Later, the food can be quickly rewarmed, using conventional means, for preparing meals that use these foods, such as tostadas or burritos. This method permits me to cook many different foods for later use. Traditional cooking methods can blend nicely with solar oven use.

What do I do when it's cloudy?

I tell people, "I cook the same way you cook on any cloudy day." All cooking methods have limitations. One great feature of solar ovens is that you don't need to buy a backup system as you must with most alternative energy tools. Almost everyone already has traditional cooking facilities.

Naturally, the solar oven needs the sun to be effective. A bright cloudless day works best, but even the first level of haze will get the food cooked. Occasional clouds passing over the oven will lengthen the cooking time slightly.

Accept the oven's limitations and have fun learning to adjust to your environment. If a forecast predicts clouds in the afternoon, then either start cooking earlier or switch to a conventional cooking method. Common sense rules.

What happens if it clouds up after I begin cooking?

This is one of the limiting realities of cooking with the sun. Without good direct sunlight, food will not cook properly. Please don't let this discount solar cooking as a viable option. Just remove the food and cook it with the other methods you presently use.

What do I cook when it's partly cloudy and it might or might not clear up?

When you look up at the sky with a real question whether the clouds are building or disappearing, you have a problem. If you think that the clouds are building you might go ahead and cook with more traditional cooking methods.

Chapter 4: Questions & Answers

If you want to solar cook anyway and there is a chance that the clouds are thinning, there are some cooking strategies you might use. You can set your cooker up and put a cooking vessel with the necessary water in it. If the sun stays out long enough to bring the water towards boiling, you can then make a judgment whether there is enough sun to fully cook the food you want. If the day has cleared up you can go ahead and prepare long cooking foods like beans. If it is still partly cloudy, some grains will cook relatively fast once the water is up to temperature. Also, you can always try to cook quick-cooking foods between clouds.

It works well in Arizona, but what if I live in Canada?

Most of my work in designing ovens took place in the mountains of northern California and Oregon. I have also used solar ovens on the border of Canada and Washington quite successfully.

Many northern areas have very sunny days, and solar cookers will work wonderfully at these latitudes. Choose ovens that tilt towards the sun in the northern latitudes because they can better focus on the sun when it's lower in the sky. In the summer, areas of the North have longer days, but in the winter the hours for cooking will be very short.

How do I focus the solar oven?

Do not look at the sun to aim your cooker. This is a natural but bad habit. Over time this habit could cause eye damage. Learn to watch the shadows created by your cooker. The sun, being very far away, creates sunlight that reaches us in parallel rays. If the shadows are even on all sides, the cooker is directly focused. With a little experience, you will be able to use the shadows created by the oven to fine-tune the focus. Learn to set the oven "ahead" of the sun's path so that the sun crosses over your cooker. By focusing the cooker slightly ahead of the sun you can learn to control your cooking times and temperatures better.

How often should I refocus it?

How often you focus the solar cooker will depend on your cooker design. With the SunStar you may wish to refocus every hour or so to keep your cooker at maximum power. This will keep the foods cooking at an even pace.

After you use your cooker for a while you may notice that in different areas of the sun's path, there will be a slightly different period of time that your cooker seems focused. At noon

Above: Center the oven in its own shadow to focus directly on the sun.

with the sun high in the sky, it moves more quickly past the focus of your solar cooker, and you'll need to refocus more often. As the sun is about halfway to sunset it seems to hang for a while, and you won't have to focus as often.

Most solar cookers will direct nearly 100% of the light through the window of the ovenbox, even while not being perfectly focused. The cooker will be able to cook at top efficiency over some period of time, depending on the design.

Many meals cook without this focusing. For example, you can put food in the oven, set it for the early afternoon sun and come home to a cooked meal without ever having to refocus. As the sun moves into direct alignment with the ovenbox it slowly climbs to cooking temperature. As the sun passes by the direct focus point it only has to add enough light to the box to offset the heat lost through the glass and insulated box. Thus, cooking takes place long past its optimum focus.

What kind of cookware can I use?

Dark pots with lids work best. Dark pots change the light from the sun into heat energy. Lids are important because

Chapter 4: Questions & Answers

steam will dissipate much of the heat (cooking power). Cast iron pots, because they are black, seem to be natural solar cooking pots. However, it takes just as much energy to raise the temperature of a pound of food as a pound of cast iron. Much of the energy coming into the oven goes into heating the pot rather than the food.

Metal pots can be painted black. You can also use enameled steel pots, usually sold in blue or black. These work nicely because they are relatively light weight.

I like glass pots. They are very clean and you can see the food cooking. If you use black pots you have to open up the oven and handle hot pots to check the food. Valuable heat is lost. Glass casserole dishes with lids work very well, especially brown tinted ones.

My favorite cooking vessels are recycled jars with lids. These can be useful in sizes 12-64 fl. oz. Gallon jars are too big and the glass is relatively thin. Sometimes they will crack. Often I paint the outside of my cooking jars black.

Left: Re-using old jars as cooking pots is inexpensive and works great.

For baking breads, cookies, pies, etc., use naturally darkened baking tins or cookie sheets, or you can buy new black bakeware.

Can solar cookers work without the reflectors?

An empty SunStar without reflectors may reach 180° F (82° C) or more. This heat is sufficient for cooking foods requiring lower temperatures. Sprouted wheat breads like to be cooked at 120° F over three or four hours, so you can cook them without reflectors. I've also used the ovenbox alone to make yogurt, watching that it doesn't get too hot. On cold days, I've used the box without reflectors to raise bread.

Heaven's Flame

Can I preserve foods with a solar cooker?

Yes, a solar cooker can be used to can fruits and tomatoes. Tomatoes must not be over ripened. Do not can other types of foods. These are not pressure cookers and there will be the danger of botulism if you try to can anything but acid fruits and tomatoes. Please read up on canning before attempting it. But don't let this warning stop you from canning. It is a very special way to use your cooker. It is especially useful if you only have a small amount of fruit to can at one time, say one to four quarts.

Are there other uses for solar cookers?

Yes. They have been used for disinfecting medical equipment where fuel isn't available. They can also be effectively used to pasteurize water. Water-born disease organisms are killed at 151° F (66° C). Solar cookers have other uses, such as separating beeswax from honey. I've also made candles in my solar cookers.

If they work so well, why aren't they widely used?

"It's hard for people to change something so traditional as cooking methods." A few years ago no one had heard of microwave ovens. With proper promotion and education, a demand could be created for solar ovens.

"Most people work during the day." Solar ovens can cook a very pleasing meal alone so you can come home to a hot meal.

"Solar energy is not reliable enough." Many regions of the world get sunshine practically every day of the year and yet don't have a solar oven industry.

"They can bake cookies, maybe." Solar ovens are workhorses and can cook virtually anything.

"It's because there's no money in it." Although solar energy is free, there is plenty of money in developing the solar industry.

Every household is a potential user. There is a market for an entire range of designs, from cardboard to high tech, from utilitarian to beautiful. A budding industry could support many people teaching others how to use them.

Chapter 4: Questions & Answers

I believe the real reason solar cookers are not widely used is because their potential has simply been overlooked.

It's a really good idea, but I can't use it in my lifestyle.
Solar cookers, with a little effort, can fit in a wide range of lifestyles. People who use their lifestyle as an excuse make me wonder about the nature of the lifestyles themselves. Lifestyles are often seen as a something that we have no control over, limiting us to the status quo. This is really not the case. We all create our lifestyles by a series of choices we make in response to different circumstances. Our choices are based on the assumptions we hold about life itself. In a world of changing forces, we must continually reevaluate these assumptions. All of us are free to change. It only entails making different choices and stepping in a new direction. Solar cooking offers us a simple way to step onto a new path.

Heaven's Flame

Chapter 5

The Work of Others: Designers & Promoters

I've known for a while that I would be rewriting this book. The last edition came off the top of my head, but to prepare for this version, I had time to write letters to people all over the world. Most of these people I had met at various solar gatherings. Others I heard about through the solar grapevine.

In this chapter I'll share with you some of the information people have sent me. This information will let you see inside a small but growing worldwide movement. All of the solar cooking designs available today are more powerful if standing together as a collection. People on Earth have different cooking needs. There is no way one solar cooker can fill all of them.

I received a wide range of material, from the very detailed to the much more general. So as not to get bogged down in details, I will give you a glimpse of each cooker. If you want a detailed look at a particular design, it is best to go directly to the people working with it. They are the most aware of their own cookers, intimately knowing the strengths and limitations of their designs.

Instead of creating a centralized information distribution system for this grassroots movement, we are creating more of an interlocking web of information sources. This is important because most of the designs are still evolving, and only those working with the particular designs are up to date. If you want to educate people about a particular solar cooker, it is best to share information that reflects the newest work on that cooker.

Heaven's Flame

The addresses of the people and organizations discussed in this chapter are in the back of the book. Right now, the movement is small enough so that most of us enjoy hearing from others who are interested in what we are doing. I encourage you to continue seeking out information on solar cookers. There is a lot of good info, and more coming every day.

Solar cooking is an expanding field. The following is an overview of what's happening in the solar cooking world. This collection shows a good cross section of traditional designs and suggests the potential of many more designs still to come.

It seems logical to divide the efforts of both designers and promoters into two different chapters, but in the world of solar cookers this just does not work. There are no pure roles of designing or promoting at this point. Once a person comes up with a design they find themselves facing the much larger task of getting it out to the people who need it. People promoting a certain design often find themselves slightly redesigning it to make it easier for them to use or build in the location that they want to serve. In this chapter, designers and promoters will be intermixed to best tell the solar cooking story.

Each designer has, from their own experience, defined what they feel are important goals in the creation of a useful solar cooker. They have followed through, designing cookers that meet these individual goals.

In the 50's, many of the basic designs were already defined. It is often the simple additions that create the greatest breakthroughs, leading to the increased acceptance of these solar tools. When you look closely at these designs, you can see that they have matured, becoming very useful cookers. We are, as a group, ready for serious attention by the general public. I will start with the simplest solar cookers and move in the direction of ascending cooking power.

Solar Panel Cooker (or the Salad Bowl Cooker)

This very simple solar cooker is fairly new, thus it is still evolving. The Panel Cooker comes from a solar pioneer in

Chapter 5: Designers & Promoters

Left: Roger Bernard's "Salad Bowl Cooker" is simple, effective, and inexpensive. An overturned glass bowl creates the oven space.

France, Roger Bernard. Roger is the author of an interesting book on solar cookers, *Le Soleil a Voture Table*, published by Edition Science from Lyon, France.

He found that if you made a few simple cuts in a cardboard box, then apply foil to the cardboard, the resulting panels can be arranged to reflect sunlight towards a center cooking area. Here, he places a small pot with food and covers it with an upside down, clear glass salad bowl. Thus his name for it, "The Salad Bowl Cooker." He found it could slowly cook useful quantities of food. Roger found it interesting that such a simple design could cook at all. He shared this cooker with others because he was having fun using it.

When Barbara Kerr, one of the designers of the Solar Box Cooker, saw it, something clicked. She thought it was potentially the cooker that many people had been looking for. For years, people at Solar Cookers International (SCI) had faced the fact that in many areas that need solar cookers, glass is not available or is expensive. SCI was always looking for a simple cooker that did not need glass glazing. Barbara starting using oven bags in place of the salad bowl. The bags can take the heat if they are slightly inflated so that the plastic doesn't touch the hot pot. She renamed it "The Panel Cooker."

She enlisted many interested people from SCI to try it out. There were diverse reports, from not cooking at all, to cooking one meal faster than the solar box cooker these people were using.

New variations of the Panel Cooker are being tried in different areas. This cooker should continue to evolve. See the picture of Roger's new addition to this design. He continues to use the salad bowl. Oven bags are not available in France or many other areas of the world. The glass salad bowls are fairly cheap and can be used for other purposes.

It is in emergency situations and refugee camps where the importance of a cooker like the Panel Cooker will make itself felt. Thousands of them could be shipped on a couple of pallets. The panel cooker allows the people to cook a small family meal. The ability to cook a meal instead of standing in line would be a first step of stabilization for families try-

Above and left: The "Cookit," Solar Cookers International is promoting these simple and inexpensive cookers in East African refugee camps.

Chapter 5: Designers & Promoters

ing to regroup. This cooker, by being simple, enlarges the whole potential of solar cookers.

Solar Cookers International has recently refined this cooker, making it extremely easy to build and use. They have also started educational projects with this cooker in the Kakuma and Dadaab refugee camps in northwest Kenya. Volunteers have made remarkable achievements in spreading the use of the new Panel Cooker. This is fast becoming one of the most interesting and exciting stories in the solar cooking world. Getting useful cookers to people, and having them accepted, has always been a prime goal of all solar cooking advocates.

The Sunstove

The Sunstove is another relative newcomer to the collection of solar cookers that is already making a significant impact. Richard Wareham started playing around with this design about four years ago, though his involvement with solar goes back much further. Living half of the last 28 years in southern Africa, he became intimately aware of the need for solar cookers in areas with severe shortages of firewood. Richard realized that for solar cookers to fulfill the potential to reduce environmental degradation, we need to get them out in greater numbers.

From his personal use of solar cookers, Richard understood that efficiency could not be the prime goal, but one that he would maximize as best he could. If he designed his cook-

Above: Richard Wareham's "Sunstove."

er with efficiency as his prime goal, the cost would be beyond the means of the people for whom he was designing. Instead of high efficiency, his goal was to make a cooker that would cook a useful amount of food for a family. It also had to look nice and be easy to use. He was also looking for one that would be easy to mass produce in a village or factory setting. Ruggedness and cost are also factors he considered.

What he came up with was basically an inverted cone with a slight angle on it. The insides of the sidewalls are aluminum to reflect light onto the cooking pots. The sidewalls are insulated. The cooking area is enclosed with a sheet of clear, unbreakable, UV resistant plastic. By not using exterior reflectors, these cookers are very resistant to being tipped over by the wind. Their conical shape allows stacking one on top of another for easy shipping. The cooking space is designed to hold two cooking pots. This cooker is not a fast cooker, but creates nice meals for a family. This is what is needed.

The Sunstove is simple to use. Richard found that, in many cultures, it was difficult to explain how to focus cookers. This was especially true with those cookers that needed to be pointed directly at the sun. With the Sunstove, Richard used an angle that allows it to cook without complicated focusing techniques. Focusing is accomplished by just pointing it in front of the sun, with a touch of the foot. The cooker does not tilt toward the sun, which allows it to be built without an adjustable rack, again simplifying the design.

The Sunstove can be built in a factory or in a village setting. For the last couple of years, Richard has been producing these for sale in Africa at a rate of about 150 per month. The steady rate shows consumer acceptance. Sunstoves are being used and sold in South Africa, Kenya, Botswana, Mozambique, and Mexico. I heard that the South African government and other funding agencies are now ordering Sunstoves one thousand at a time. The National Association of Congregational Churches, to which Richard belongs, is setting up a small production line in Mexico to build these cookers. It looks like this cooker is one of the successful stories in solar cooking.

Chapter 5: Designers & Promoters

Richard has patented this design, but offers it freely to non-profit groups. He can be hired through his own non-profit group to help set up production lines and to promote these cookers. Currently, the cost is about $25 each.

Solar Box Cooker

Worldwide, this style is probably the most common cooker. It is a large insulated box with a glass top and one reflector that can fold down on the glass or open up to reflect extra sunlight. This glass should be at least 4 square feet for effective cooking. It can be larger, which will result in more cooking power. This cooker lends itself to being built out of many different materials such as wood, metal, cardboard, or even mud and brick.

The design is adaptable. Some people have even built them into the south facing walls of kitchens so it can be used from inside the house. In Switzerland, Gruppe ULOG puts a slight angle on the glass so it works better in northern climates, but uses level glass when building them in the trop-

Left: A typical, single reflector solar box cooker.

ics. Other people build them on a stand with wheels to make them easier to move around. I've even seen them double as a patio table.

This cooker works surprisingly well for slow cooking. The Solar Box Cooker also lends itself well to cooking while you are away. The oven box is quite large. It can hold more than one pot, enabling you to cook a multi-course meal. This cooker does not tilt towards the sun, so no complicated rack system is necessary. It only needs a rack to keep the pots off the bottom so heat can circulate. In the northern climates, this cooker is limited to the months having longer days, but in the tropics this cooker can be used all year-round.

The Solar Box Cooker is in use all over the world. Both the Chinese and the Indian governments subsidize them. China probably has more of these cookers in use than the rest of the world put together. India has even developed national standards for this style of cooker. It is produced by many companies with various subsidies. Solar Box Cookers are now being promoted in Africa, Central and South America, Europe, and the United States.

There are quite a few groups promoting this style of cooker. Here are just a few of them:

Solar Cookers International (SCI)
The reason many of us are aware of the Solar Box Cooker stems from the work of two women in Arizona, Barbara Kerr and Sherry Cole. During the 70's, many of us were becoming aware of alternative energies. Barbara and Sherry became interested in the potential of solar cookers. They started using Sam Erwin's Solar Chef and Dan Halacy's 30-60 design, both described later, for potlucks with friends. Then Barbara heard of a design from India for the Solar Box Cooker. Before she found out they that were meant to be built from wood or metal, she built one from cardboard, which worked surprisingly well. The Solar Box Cooker, because of its ease of use, was soon to be her favorite.

Barbara told me an interesting story from the early days of sharing this tool. She invited an engineer from the nearby

Chapter 5: Designers & Promoters

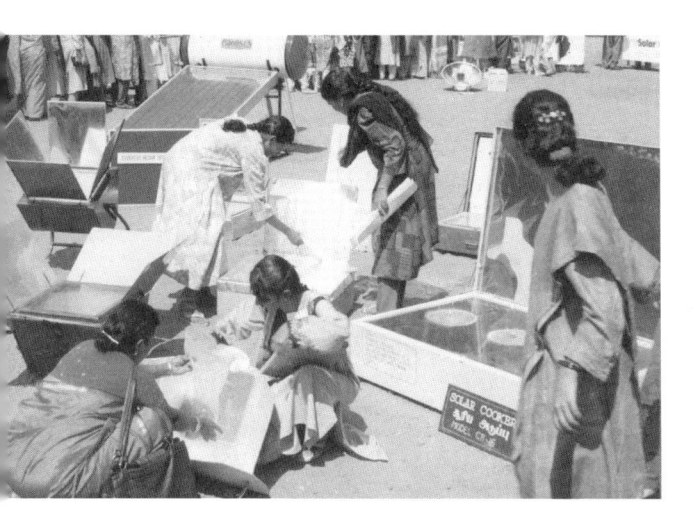

Left: SCI at the third International Solar Cooking Conference in Coimbatore, India, January 1997.

university to a solar potluck. While the boxes were quietly cooking the afternoon meal, the engineer proceeded to tell her why the Solar Box Cooker would not work, not even noticing that it was sitting there cooking right next to him.

These two women, fortified with many solar meals, went on to produce the Eco-Cooker out of cardboard and offered it for sale. These cookers got into the hands of other people who were also impressed, including Bob Metcalf and Bev Blum, from the Sacramento California area. These two, along with others, decided this tool needed a non-profit group to promote it to the world. Thus, Solar Box Cookers International (SBCI) was born. Since then, as their role as promoters has evolved, they have become Solar Cookers International.

Bob has shown these tools all over Africa and the Americas. Many of these trips were partially funded by Pillsbury. I'm sure they did this for humanitarian reasons, but I've got to think that they also realized that if solar cookers became widespread, they would have a whole new market for their baking products. This involvement with large corporations can help greatly in getting these tools out, and we are grateful for Pillsbury's effort.

Bob Metcalf, a professor of Microbiology, is primarily responsible for recognizing the potential of solar cookers and solar energy to pasteurize biologically polluted waters. As he traveled the world with the Solar Box Cooker, he became intimately aware of water problems facing many people in developing nations. Water-born organisms are the number one killer of children under five, but can easily be eliminated with a solar cooker. He found many people willing to use their Solar Box Cookers to pasteurize water, even if they did not want to use them for cooking.

SCI continues to promote Solar Box Cookers. They send out plans, teacher aids, and newsletters all over the world to keep the interest in solar cookers alive and growing.

Lately, they have been sponsoring the World Conferences on Solar Cooking. There have been three: in California, Costa Rica, and India. They have also co-sponsored regional conferences in Honduras and recently in Kenya and Ecuador. These conferences are proving to be a real boon to the solar cooking movement, allowing promoters and designers a place to talk shop and share experiences. As more groups with projects that could benefit from solar cooking become aware of conferences, and start attending, the importance of these conferences will grow.

Gruppe ULOG

There are some other groups that are responsible for the widespread use of the Solar Box Cooker. One is Gruppe ULOG of Switzerland and Germany. This group has its origins in Botswana, where Lisa and Ulrich Olehler share these cookers with the local population. Their early experiences were that many people were resistant to solar cooking. They felt that this was due to the locals seeing solar cookers as second class tools, because people in the developed world don't use them. This resistance has been noted since the very first days of promoting solar cookers.

It may seem like weird reasoning when you look at the need for cookers in these areas, but we are dealing with human nature. The developed world tells these people through an overwhelming amount of advertising that they should follow our lead into a consumer society. A world where the amount and kinds of things you have gives you status and fulfillment in life. Our own people have accept-

Chapter 5: Designers & Promoters

Left: A beautiful Gruppe ULOG box cooker in Sudan.

Below: The basket cooker utilizes traditional weaving techniques as an alternative to expensive wood building materials.

ed this barrage of advertisement, so why do we believe it should be easier for people in less developed areas to see the fallacy in this world view? When we try to educate people about solar cooking, we need to understand human nature and work with it.

Ulrich decided that if he was going to succeed in sharing these tools, he had to start at home. On his return to Switzerland, he and Lisa shared them with friends and started a solar group that later became Gruppe ULOG. Their goal is to promote the use of solar cookers at home and abroad. Since the early 80's, this group has prompted over 5000 households in Switzerland to integrate solar cooking into their lives. From this base, they are once again reaching out to successfully promote cookers in Senegal, Kenya, and Sudan. They are now finding women quickly adapting to solar cooking. Now, the oven building projects they start are quickly being taken over by locals.

SERVE

Another group promoting solar box cookers is SERVE, a Christian relief agency working in Pakistan with Afghan refugees. They needed a rugged, good looking, and inexpensive solar cooker that could cook for a family of 8-9 people. They had information on Solar Box Cookers, and like many promoters, redesigned it to fit their needs.

At first they built with wood but then used flute board, a plastic cardboard that comes in a variety of colors and is produced in Asia and the United States. SERVE used this material on the outside because it's lightweight, attractive, fairly inexpensive, waterproof, strong, and has good insulation properties. They also were able build their cookers without using a hinge. Not needing a hinge may seem like a small point, but it's these kinds of materials that cause the cost of a solar cooker to add up. Unfortunately, they had to import the fluteboard through customs, causing many delays. They eventually abandoned the material because of this and because it was a little harder to work with than they anticipated. Now they are back to building out of wood.

Since the end of the war in Afghanistan, SERVE has moved their promotion effort to Kabul. They brought in 750 solar cookers which sold immediately. The refugees returning from Pakistan brought with them the awareness of solar cookers. Kabul was still without reliable electricity, so the cookers were sought after. Many more are needed.

SERVE distributed over 5000 cookers to refugees in Pakistan and found them well accepted. The United Nations subsidized the cost of these tools. The practical experiences of this group should be recognized by others involved with refugee situations. SERVE is willing to share this experience.

I attended a talk about this project at the World Conference. The spokesperson made an important comment. People were willing to try solar cooking, except in areas where other solar cookers that did not work well had been shown earlier. This created a barrier that was hard to over-

Chapter 5: Designers & Promoters

come. If you want to share solar cookers, it's important that you learn how to use them first. You need to convey a sense of confidence in what your cooker can do. Know its strengths and limitations and be honest with what it can do. It is better to have no project than an ill-conceived one. This isn't hard: use the cooker you want to share in your own life, and learn how to share this knowledge with others.

Solar Energy International

Solar Energy International (SEI) is a non-profit organization dedicated to promoting the use of renewable energy technologies through education and technical assistance. SEI offers hands-on workshops in solar electricity, wind power, micro-hydro power, solar home design, environmental building technologies and solar cooking. One of their programs promotes solar cooking around the world. At their facility in Carbondale, Colorado they offer three day workshops on how to build and use solar ovens. Each year, SEI holds an annual Solar Potluck and Exhibition to share solar cooker technology and delicious solar cooked

Below: SEI designed and taught the building of two of these commercial sized box cookers for a women-owned bakery in Sonora, Mexico.

food. Staff member Ed Eaton was one of the founders of the Tucson Solar Potluck and Exhibition which is now in its 14th year. SEI also works internationally to promote solar cookers.

In 1991, SEI staff member Laurie Stone went to El Salvador and taught women in a rural repatriated refugee community how to build solar ovens. These ovens were modeled after Bill Lankford's Central American Solar Energy Project. Solar Energy International also worked in Sonora, Mexico to help women in an impoverished neighborhood start a solar bakery. SEI provided the technical assistance to help the women build two large commercial size solar ovens designed by Ed Eaton. The women of the solar bakery now bake and sell hundreds of cookies, cupcakes and empanadas weekly. The solar oven project provides them with a salary, while helping to spread the word about solar cooking.

Avinashilingam Institute for Home Science & Higher Education for Women

This institute has been promoting solar cooker usage in India since the mid-eighties. They hosted the third World Conference on Solar Cooking in January of 1997.

The school also played an important role in the All India Coordinated Project on Cook Stoves, which included the popularization of solar cookers. The project included experimental studies of these cookers, motivating and educating rural families in their use. They introduced cookers into selected households and evaluated them in home situations.

They have also studied the nutritional value of solar cooked foods. This group has a lot of practical knowledge on how to create and run a community education project on solar cookers. They are willing to share this experience with similar groups. They have classes in solar cooking and are able to create group involvement in exploring the use and promotion of these tools.

These are very good people who enjoy community work. There is a need for many other community centers world-

Chapter 5: Designers & Promoters

Left: Women inspect the Avinashilingam Institute's box cooker in Tamil Nadu, India.

Below: Multiple pots cooking simultaneously.

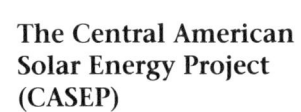

wide to take on such educational projects. Please contact them if you are trying to set up a similar community effort. This sharing can create many work opportunities for people to teach their peers solar cooking habits.

The Central American Solar Energy Project (CASEP)

This non-profit group was started by Bill Lankford to share solar cookers in Central America. Mr. Lankford has a novel method to increase acceptance of solar cookers. He starts with a questionnaire for people interested in solar cookers. From this questionnaire he can make a preliminary assessment of the probability a person can successfully learn to use Solar Box Cookers. An example question might be, "Do you have a sunny place in the yard where the solar cooker could work?" Or maybe, "Will there be anyone at home during the day to focus the cooker?"

If there is likelihood of success the person signs an agreement. In exchange for attending a workshop, in which the person builds their own cooker, the person agrees to help instruct another workshop. They also agree to return the cooker if they no longer use it so it can be recycled to other interested people.

He found people need some basic follow-up work after they receive a solar oven. If a new owner experiences a failure, or even a slightly less than satisfactory meal, this will usually be enough for them to give-up. Central Americans of all economic levels have very high standards of food preparations.

Bill hires part-time helpers from his best students to do follow-up work. They visit new owners to see if individual help is needed in learning to use the cooker. At the same time, the person checks the cooker itself to see if it needs any maintenance or repairs. If it does, they send out another hired person to make these repairs. By hiring people for these follow-ups Bill creates jobs while increasing the success rate of his work. The increased success rate more than justifies the added costs of this method. This program also creates interesting community dynamics. People who have learned how to use solar cookers are empowered to help their neighbors make the solar transition.

Bill has worked on this project long enough to notice some of the social changes that are altering the lives of the poor people who are his prime solar candidates. "Structural adjustments" imposed by large international funding agencies (IMF, AID, World Bank, etc.) create many hardships and disruptions of traditional cultures and individual families. Layoffs or salary reductions of the male heads of household and reduced spending for social services such as medical care, transportation, and education bring survival crises to many families. The last employable member of the family, the mother, is forced to accept outside employment. The mother is usually the person who learned how to use the solar cooker. Bill has started teaching older children in the family who are still at home how to use the solar cooker.

Chapter 5: Designers & Promoters

The cookers that he builds cost about $150, which is beyond the means of most of the families that need them. Usually, his cookers are subsidized to these families. He feels that their effort to build cookers and to help with future workshops is already quite an investment in time and effort to those living on a survival level. Finding funding to help families with the initial cost of cookers has become difficult because of the new emphasis on free market economic philosophies behind these traditional funding sources. He has found a few sources that will still provide money for humanitarian reasons, but they want to see a high level of acceptance of the cookers. This is why projects like his are so valuable. Bill's work on this local level has very important ramifications for many areas of the world.

CASEP started in Guatemala and has spread to five Central American nations. In Costa Rica some women started their own group based on this effort called "Sol de Vida," (Sun of Life). In Honduras, another group of women started "Defensoras de Vida," (Defenders of Life). Defenders of Life is a pretty powerful name for a group of women promoting solar cookers. These names show how important the promotion of solar cookers is to these women and their communities.

Sunlight Works

Bob Larson and Heather Gurley, two long time solar advocates from Sedona, Arizona, created Sunlight Works as a vehicle to educate and promote solar cooking. They have 25 years of experience teaching about passive solar tools. During these years, they have directed 45 community projects in solar cooking in rural and inner-city communities. One recent summer, they solar-catered a meeting of 200 scientists in Colorado. They have been hired to teach solar cooking by the Arizona Energy Commission (another long time promoter of solar cooking). Bob and Heather have helped to keep solar cooking in the public view for a long period of time.

Heather has written a science workbook for kids called *SunLight Works*. I gave a copy of the workbook to a teacher I

Above: A Sunlight Works fundraiser for the Sedona, CA Boys & Girls Clubs.

know. She loved it and is now taking her class through it. Teaching kids about solar cooking is extremely important. They are our future mothers and fathers. Heather also created a nice solar cookbook, *Solar Cooking Naturally*, which I use. Writing solar cookbooks is a field that will expand as the number of solar cooks increases.

Bob Larson's work centers on consultation and training. His background includes the Peace Corps in Nigeria. He worked with the Solar Box Cooker, among many other designs, and is available for consultation on solar cooking projects. Bob and Heather can contribute valuable experience to any group wanting to promote solar cooking.

Multi-Reflector Solar Cooker

Multi-reflector cookers traditionally have four reflectors that angle out at 60 degrees from the glass. They usually have a small ovenbox relative to the amount of sunlight their reflectors intercept, so temperatures are hotter than previously mentioned cookers.

This style played a prominent role in the 1950's studies of solar cookers. From the view point of efficiency, size, and power, this is how solar cookers should be built. However, many people directly blame these cookers for the failed attempts at introducing solar cooking. They were seen as too expensive, complicated, and awkward. They were also susceptible to tipping over because the reflectors are arranged in a way that catches the wind. Their cooking area would often hold only one pot, while most meals consist of more than one cooked dish. All these problems created negative experiences with this style of cooker.

Chapter 5: Designers & Promoters

Multi-reflector cookers usually have 100° F more power than the commonly used simpler styles. For many people, this makes this cooker very attractive, even with the acknowledged limitations. Many designers, myself included, decided to work with this style, seeking to minimize the traditional problems. My style, the SunStar, falls into this group. Here are some of the other people who have worked with this style.

Bud Clevett

Bud's involvement with solar cookers goes back to the early 1950's. He was an archetype of the creative inventor, working with the likes of Bucky Fuller. Bud was an inventor with over 60 patents to his name, some of which were used by NASA. I saw one of his patents for a lightweight, strong, and foldable bridge that was a stroke of simple genius. He held workshops throughout the USA on the creative process. This shows the caliber of the people that first recognized the potential of solar cookers. This solar tool caught Bud's imagination and became a life-long hobby.

Over the years, Bud has designed over 50 different cookers. The times in which he lived were reflected in his designs, some of which were true 50's art deco pieces. One of them was called the AstroOven. It looked like it belonged on the set of some science fiction film. Another was called the Sunflower. This patented cooker was complete with petal shaped reflectors, a stem, decorative leaves, and was planted in a flower pot. Others included the HelioOven, the Coconut, and a more traditional Solar Range. His family says we may see some of these designs marketed in the future.

Right: Kathleen with "Sunspots" both opened and closed.

53

Heaven's Flame

The design he is probably best known for is the Sunspot. The Sunspot is a small cardboard backpacker type model still being manufactured. They are popular as educational tools for teachers. He started selling these in the early 60's and sold an average of 1000 per year. This is a remarkably large number for those early years.

The Sunspot is lightweight, weighing only 1.5 pounds. It is also very small, with a cooking area of 4 x 10 x 10 inches. The Sunspot cooks small meals, reaching temperatures near 350° F.

Bud recently passed away. Though he is gone, his contribution to the world of solar cookers will continue to be felt. In his death, we are reminded of the many people who have contributed to the promotion of solar cookers. Goodbye, Bud, your work is appreciated.

Dan Halacy

Dan was also one of the early solar pioneers. He attended the First World Symposium on Solar Energy in 1955, and wrote over 70 books on solar energy. His most popular book, *Solar Science Projects,* was geared towards children in school. Scholastic Press sold over three hundred thousand copies. It stimulated many young minds. In the 70's Dan wrote and published a very good book on solar cooking. *Cooking with the Sun* includes plans for a cooker he designed. The ovenbox of his cooker is made so two different sides could become the bottom of the cooker, resulting in a choice of two different glass angles, 30 and 60 degrees. This allows the cooker to focus when the sun is

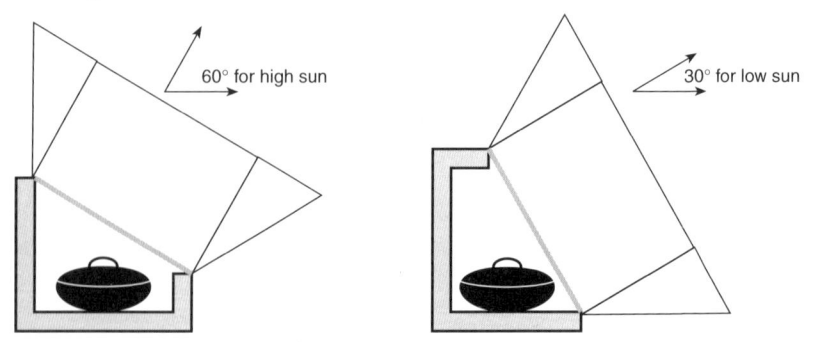

Above: Dan Halacy's 30°/60° cooker design is versatile at any latitude.

Chapter 5: Designers & Promoters

Above: Beth and Dan's solar cookbook.

Cooking with the Sun

How to Build and Use Solar Cookers
- Simple solar cookers that can reach temperatures up to 400°
- Enjoy the higher nutritional value of solar cooked food
- No smoke, no ashes, Just clean heat

Beth Halacy and Dan Halacy

either high or low in the sky. The book describes how to build this cooker using a minimum of wood.

The book also contains an extensive cookbook section. This cookbook is unique in how it teaches people to use the cooker. There are symbols depicting different types of solar days (full sun, hazy days, cloudy days, etc.) included with each recipe. This helps new solar cooks decide what to cook on different days. This book was re-released in the early 90's, so you might still be able to find a copy.

When I attended the Tucson Annual Solar Potluck a few years back, the 30/60 degree style solar cooker was the most common style being used there. Many people were very happy with it, and used it regularly.

Burns Milwaukee's Sun Oven

Tom Burns is the designer and founder of this company. He succeeded in producing a very nice cooker for sale to the general public. When I used to daydream about various ways to work with solar cookers, I'd dream of building a cooker that looked a lot like this one. For me, this was just a pipe dream, because I knew I did not have the business and manufacturing skills necessary to pull it off. I was so glad to see the Sun Oven. It is an important addition to solar cooking, and is so clean looking it could be sold through a wide range of stores, especially any selling outdoor cooking equipment like barbecues.

The body of this solar cooker is a nice off-white fiberglass. The window is made of tempered glass which is very strong and safe. The tempered glass is fitted on the oven

Above: Tom Burns explains the "Sun Oven" (left) and community sized "Villager" (right) to a future solar cook.

box in a hinged wooden frame. The glass is used as the door. Tom developed a unique reflector system that makes it easy to set up and break down. The reflectors are made out of spectral quality aluminum. A gimbal rack system is included, so the food stays level when the cooker is tilted towards the sun. The cooker weighs a little over 20 pounds, and as a total package is a nice consumer product.

Burns Milwaukee has produced a larger version called the Villager. It can cook 50 loaves of bread per hour, or hundreds of loaves each day. It is so large, some are sold with a trailer. The Villager has an option for a built in propane back-up. If a baker lives in an area where clouds often build up unexpectedly, he can quickly shift to a back-up power source. With the insulation required for solar cooking, these end up being very efficient propane cookers.

Chapter 5: Designers & Promoters

This company actively promotes their cookers, especially in Jamaica. Burns Milwaukee donated a Villager cooker to the St. Patrick Foundation in Kingston. I wrote to this foundation to find out what they were doing with it. They wrote a nice letter saying they were putting it to good use, and hoped they could get a few more. A baker gets up early in the morning and starts his dough so that at first light he is ready to start creating a stream of baked goods to offer to the community. The solar bakery helps support their schools and other educational programs.

These cookers are expensive, compared to some other cookers. The cost could come down considerably, if they were built in the third world country where they were to be sold. Many people in these countries enjoy Tom's effort and appreciate the quality and looks of the Sun Oven. The durability of this cooker makes it important in areas with no infrastructure to provide the follow-up maintenance required on many cheaper models. There are very few problems that can develop with this cooker.

Tom deserves a lot of support for his work. For many years, people in the solar world felt that it was important to show that solar cookers can succeed in a consumer market place. I believe he has succeeded.

Blazing a new trail is much harder than following others. His company not only has to compete with other solar cookers, but also has to spend time and money educating their potential consumers that solar cooking is possible. As our collective education efforts raise the awareness of solar cooking in the general public, more advertisement money can be aimed at sales. This will help their bottom line. I want to personally thank Tom Burns for this fine pioneering effort.

Joe Froese

Joe is a good example how and why solar cookers are still around. He saw his first cooker in the early 80's, and something clicked. Here was a simple tool he could use to do something meaningful, helping both people and the environment. In his own backyard he began to design and use cookers. As his success in building good cookers grew, so

Heaven's Flame

did his confidence in them and their potential. At the same time that he was refining his cookers, he was also honing his visions of ways to promote them.

Joe's home is in Saskatoon in north central Canada. Building and using solar cookers in a cold, sunny, northern climate is a really good way to become impressed with them. In the North, cookers that tilt towards the sun and have multiple reflectors become almost essential. Joe developed a line of Freedom Cookers that work well in northern climates, and therefore also work great in extreme southern areas. These come in large family size or larger institutional models.

One of his first efforts to share solar cookers on the world stage was when he organized a trip to the newly independent African country of Eritrea. One day, while building a solar cooker, a few people gathered around him to check out what he was doing. As the day progressed, the

Above and left: Joe Froese's "Freedom Cooker" tilts to follow the sun and has multiple reflectors, important features in extreme latitudes.

58

Chapter 5: Designers & Promoters

numbers swelled to over a thousand. Drums came out and songs were made up about how the sun was going to help their lives. A truly spontaneous solar celebration took place. This is an area with little cooking fuel and the people could easily see the value of these tools. They've communicated to Joe their desire for these tools, but they lack the simple materials in their country to build them. This is one more story showing that people are ready to try solar, but the cookers are not available.

Through teaching and building cookers, Joe has reached out to people in northern Africa, Mexico, Haiti, China, Fiji, Burkina Faso, and more. Joe's new effort is to share his cookers with Cuba, where he has been spending winters. I heard he married someone he met there. Maybe it's one of the first solar cooking love stories. All this started in a backyard in Canada.

Sam Erwin's Solar Chef

The Solar Chef is a unique design. It's discussed here with the multi-reflector cookers, although, in a traditional sense, this design deserves its own category. It is a blend of the multi-reflector cookers and the parabolic cookers. This accounts for the higher temperatures of this design.

I first saw this cooker in the late 70's, cooking at an alternative energy fair. It was a very beautiful cooker that reached temperatures well above 400 degrees F. It duly impressed many people at the fair by cooking large quantities of food.

Hotter than most of the multi-panel cookers, this one cooks food in about the same time as a conventional stove top. The cooking area is an eight sided glass cone surrounded by mirrors that reflect light directly onto the food. The glass cone and the arrangement of the mirrors makes the Solar Chef a real work of art.

When I first saw it, Sam used blocks of wood to tilt this cooker towards the sun. Since then, he has greatly improved the design by making the stand very easy to focus. The stand also makes it very easy to move the cooker around the yard.

Heaven's Flame

Left: Sam Erwin's original "Solar Chef" is durable and very efficient.

Below: Sam's prototype for a new "Solar Chef" is collapsible for easier transport, and has focusing reflectors.

Sam developed this design in the 70's in Arizona. He started out trying to design an efficient water heater but found it too hot. The idea struck him that it might cook food. Partly by accident, another cooking design was born. He built and sold many throughout Arizona.

Last year in Arizona, I met someone who had two of these cookers. They planned to keep one in storage in case the other cooker wore out. Well, the backup one is still in storage, and they still use the first one regularly. These cookers are well made and last.

At a west coast solar show, I saw a large one, 8 feet across, cooking away. Sam built it 18 years earlier. I imagined how well this type of solar cooker would look in parks and camp grounds. It could be used by people at family picnics in place of the barbecues many parks have.

Chapter 5: Designers & Promoters

Mr. Erwin patented this design and is still making them on a small scale. He might be interested in talking to people who might want to produce this cooker on a larger scale.

Sam now lives in southern Oregon. I went to visit him and look at the improvements in his design. I know he has mixed feelings about the patent process. In the 70's, he spent a lot of money on legal fees to protect this cooker because it was similar to a patent given to a British officer in India a hundred years ago. He prevailed, showing his cooker was different enough to deserve a separate patent, but it was a draining experience.

Parabolic Dish Solar Cookers

This is one of the traditional solar cooking devices that were studied from the very early days. The parabolic shape of the reflector allows the sun's incoming rays to be reflected towards one point. This concentrates an area of very high heat, reaching the temperatures necessary for frying. The immediate and high heat of these cookers is quite impressive. Depending on the size, it can have the same power as any regular stove top.

There are some problems associated with this cooker that challenges designers. These cookers have to be refocused about every 15 minutes to keep the hot point on the pan.

Another problem is the potential eye damage from inadvertently receiving concentrated doses of sunshine. This is a potential problem with all cookers that use highly reflective materials, but it is especially the case with concentrative reflectors.

Another problem with parabolic cookers is that they are susceptible to being tipped over by the wind. They have been known to start fires if the focal point lands on dry grass or twigs. Most designers of these tools are aware of this problem, and have created designs that eliminate or significantly reduce fire danger.

EG Solar's SK12

This parabolic cooker was designed by engineering students at a technical college in Altolting, Germany. It was

61

created to address the traditional problems associated with this style of cooker. They used a parabolic shape with a short focal point. A short focal point means that the point where all the reflected light meets is not very far from the center of the parabolic reflector. This is done by making the parabolic shape more cupped, or less flat. The SK12 focal point falls within the shape of the parabolic reflector. This short focal point alleviates two of the problems with parabolic reflectors. First, it allows more time between refocusings. The reflected light meets at the bottom of the pan for a longer period of time. Second, the short focal point also reduces fire danger if tipped over.

They also created a strong stand with a wide base to make it wind resistant. To protect the cook from the glare, they made it easy to rotate the reflective dish out of the way. Instead of having to reach over the reflector to the food, the cook rotates the reflector, allowing access to the food from behind.

To make the SK12 even more useful, they teach people to use a "hay box" along with this cooker. They can bring food to the boiling point and then place the pot into an insulated box. The food continues to cook as it slowly cools. This frees the SK12 to cook another pot of food, letting the cooker cook a series of pots. This cooker was designed large enough to have the same power rating as

Left: the "SK-12" parabolic cooker on display in Peru.

Chapter 5: Designers & Promoters

most electric or propane range tops. They now have a larger cooker with much more power.

The people working with the SK12 are focusing on transferring the necessary building skills to third world countries. They sent the SK12 in kit form to over 50 countries. They also sent out 1400 sets of parabolic reflector panels with information on how to build the stand. Most of these went to South America, where they have found a lot of acceptance. I saw a photographic essay about one of their projects in Africa that showed that the kits could easily be assembled in a village setting. The college students and professors working with the SK12 built a worldwide promotion effort to share this cooker.

Zomeworks' Sunflash

This is a relative newcomer to the solar cooking world. The futuristic looking design was created by Steve Baer and Lui Yoder. The Sunflash is a parabolic cooker that is relatively small when compared to others. The design allows for this smaller size by increasing the power. It focuses the sun on a pot that is enclosed in a rugged Pyrex glass sphere. The air inside the sphere creates an extra insulative effect. The increased power allows the Sunflash to cook a larger quantity of food than you would expect from its smaller size. This cooker can boil, bake, or fry.

A hole in the center reflector dish, where the shadow of the cooking pot naturally falls, allows the Sunflash to have

Left: The Zomeworks' "Sunflash" with its Pyrex sphere.

less wind resistance with no loss of power. Some of the wind hitting the cooker flows through this hole.

As with the SK12, the parabolic shape of the Sunflash has a short focal point. This allows the rays to focus on a safer spot and makes the cooker more convenient because the focusing times are extended.

The Sunflash is securely mounted on a pole and manually focused. This is another well designed cooker, one of the true space-age looking solar cooking devices that will be quite a conversation piece in any yard. One can imagine a cooker of this type coming with a tracker and a small computer brain to control the cooking process.

China's Eccentric Axis Focusing Cooker

For a long time, China has recognized that cooking with the sun's power could help their people. They started very early, exploring parabolic cookers and becoming aware of their problems. In 1975, at the First Chinese National Solar Energy Congress, Liu Zude introduced the first focusing reflector with an eccentric axis, as an alternative to the traditional parabolic cooker.

The Eccentric Axis cooker has a reflector that produces the same concentrated focal point as a parabola, but without the round dish. They used two or three slightly curved reflectors that can fold together for traveling or packing away. When folded, it looks like a large suitcase. When opened, the reflectors focus the sun's rays to a point where cooking takes place. This cooker is used like a parabolic cooker, but has a much more convenient shape.

The reflector surfaces are coated with a highly reflective aluminized polyester. This is a very attractive solar cooker. They compare in power to conventional cooking methods. The designers of this cooker developed four versions.

This was what the Chinese were looking for. By the second Solar Congress in 1978, all of the concentrating cookers shown were of this style. In 1992, one of the designers, Xiping Wang, introduced this design to the world at a solar cooking conference. He stated that over 100,000 of these units were now in use in China. We must realize that the

Chapter 5: Designers & Promoters

Above: Eccentric axis cookers from the Henan Academy of Sciences, Zheng Zhou, China.

Chinese are on to something, and maybe it's time that people in other countries check out what they have to offer. Wang extended China's desire to share their designs and experiences with countries needing them.

Xiping Wang also told us of their effort to create more efficient cooking pots. Parabolic cookers traditionally use normal pots. They get heat from below but rapidly lose heat through the sides and top. To minimize this, the Chinese developed a special pot for solar cooking. It includes a double sidewall so there is an insulating air space between the hot food and the outside air.

The Solar Hybrid Kitchen

The Solar Hybrid Kitchen stretches the potential of solar cooking to new dimensions. This cooker was designed and is being promoted by Wolfgang Scheffler and Christine Lippold, of Germany. They have now joined with the efforts of Gruppe ULOG in promoting solar cookers.

The Solar Hybrid Kitchen has a seven square meter parabolic dish consisting of 47 flat surfaces with aluminized polyester film as the reflective medium. This dish uses a simple mechanical tracking system that focuses a large amount of sunlight onto a clay oven that is integrated into a building. The cooks work inside the building in the kitchen. This cooker is a hybrid because the oven also has a place to burn wood to add extra power if needed. The pots hold up to 100 liters (about 20 gallons). This cooker can also be built with solar hot plates to cook chapatis.

In 1987, the first Solar Hybrid Kitchen was built in Kenya. Since then, Gruppe ULOG has helped to build more than 60 of these kitchens in Sudan, Kenya, Cameroon, Peru, Germany, and India. In India, a local group is commercializing them with no further assistance from Gruppe ULOG.

Most of these kitchens were built for schools, some feeding up to 400 children each day. These schools have experienced a reduction in the use of firewood of up to 90%. Think about how much firewood collection effort this cooker is saving these schools, and how much less pressure is being put on the surrounding plant life. These cookers could be used in any institutional or large community setting in sunny areas. This is one of the most exciting solar cooking projects taking place today.

At first, Wolfgang and Christine were in the field installing these cookers themselves. Now, due to interest and demand, they have started training other people to install them. They are also putting a renewed effort into improving these kitchens. Most of these kitchens initially had

Left: In this smaller hybrid kitchen in Portugal, parabolic mirrors reflect the sun into the oven box. Cooking happens inside the building.

Chapter 5: Designers & Promoters

Right: This Gruppe ULOG installation in Kenya has three huge mirror arrays and provides meals for 200 people.

Below: Inside a hybrid kitchen: Notice the space on the right for the backup combustion fuel.

been funded by charity groups. I hope for the very best for Wolfgang and Christine in their efforts to share this technology.

Solar Convection Cookers

Convection cookers transfer heat by the movement of hot air past the cooking vessel. There are non-solar ovens that use convection by putting a small fan in the cooking area to stir up the air. These are very energy efficient. Some designers have incorporated this concept into solar cookers.

In the traditional solar cookers we have already discussed, there are limitations to using air currents to transfer heat to the cooking vessel. I've had friends put small photovoltaic-powered fans into solar ovenboxes. They usually ended up not using the fan because there were no significant differences in cooking power.

Heaven's Flame

The concept is a good one, but in regular solar cookers stimulating air flow does not help, and might be counter productive. Convection may increase energy transfer to the cooking vessel, but any gain is partially offset by greater heat loss through the window. The window, even if it's a thermal pane, has a very low insulative value. Moving hot air past the glass greatly increases the heat transfer out of the cooking area. To use the concept of a convection cooker, you have to design a special solar cooker.

The Convective Solar Cooker

A cooker designed by Michael Grupp of France uses a large flat plate collector that channels solar heated air up into an insulated ovenbox. The collector is placed at about a 30 degree angle to the ground. To collect extra sunlight Michael added a reflector to the flat plate collector. This reflector is positioned to reflect the maximum amount of sunlight onto the flat plate collector.

Air heats up in the flat plate collector and naturally rises into the insulated cooking area. When the air transfers its heat to the cooking pot, the temperature of the air is lowered. As more solar heated air rises out of the collector, it forces the now cooled air down the back side of the flat plate collector. At the bottom of the collector the air enters

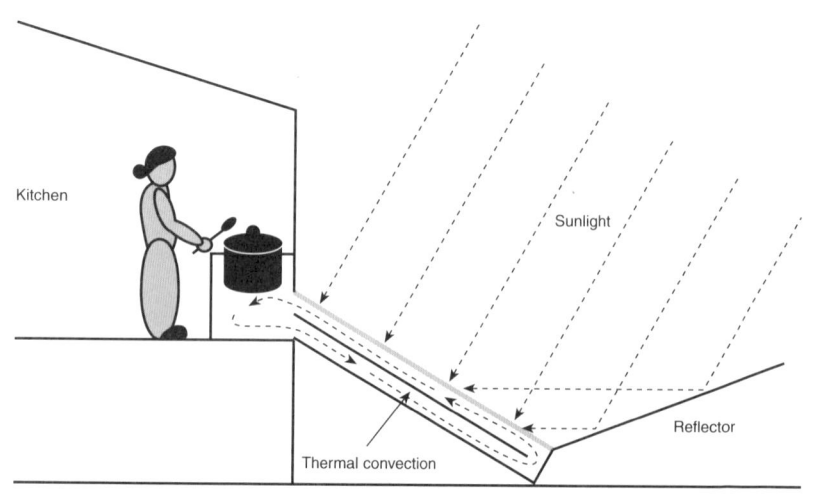

Above: Diagram of a large scale solar convection oven.

the heating area of the collector to be reheated. A natural air current, known as convection, is developed.

Metal fins have been added to the pots used with this cooker to increase their surface area. The larger surface area increases the speed in which the air can transfer its heat into the cooking vessel.

As of 1992, five institutional kitchens based on this design have been built. They were built in Cameroon, Ethiopia, Somalia, and Sudan. User acceptance was good in all cases. The larger of these cookers was able to cook up to 252 liters using seven pots. Since these cookers were installed, though, four out of five have been destroyed or damaged by war and unrest. War is definitely not good for solar cookers.

Citizens For Solar and the Tucson Solar Potluck

Citizens For Solar was founded in 1982 by Ed Eaton (now with Solar Energy International) and a group of other solar advocates who started out doing solar cooking just for the fun of it. They were inspired by a previous cookout in Phoenix sponsored by the Arizona Solar Energy Association. ASEA never had another cookout, but Ed did. In fact, Citizens For Solar has held one every year for fifteen years straight.

Right: Toby Schneider of Citizens For Solar with a demo oven and snacks.

The Tucson Potluck is the main event promoted by CFS each year, although they also do demonstrations at churches, public schools, and Earth Day events. Their main focus is to educate the public about the benefits of solar power, by actually using the equipment they talk about. Although the main focus is on solar cooking, they also have photovoltaics, hot water systems, and a whole host of other solar gadgets.

The Tucson Solar Potluck draws over a thousand people every year for a day in the park, food, fun, and to talk up solar. Most of the ovens are homemade, running the gamut of cost and performance, and showing how non-critical exact oven design is.

The Tucson Solar Potluck and Exhibition is held the second Saturday of May each year at Catalina State Park, just north of Tucson, Arizona. Similar solar cookouts are held each year in Taylor, Arizona and Carbondale, Colorado. All are as free as the sun that powers them.

Chapter 6

Designing & Building Your Own Solar Cooker

In the last chapter, we explored many different kinds of solar cookers. You saw that all solar cookers are basically very simple in design. By combining a little geometry and a little understanding of thermodynamics, a wide variety of cookers can be built. The ease of building good cookers is one of the attractive most aspects of solar cooking.

In this chapter, I will share some insights into building and designing solar cookers. It is very important to understand the design concepts. If you want to share a particular design with others, you might find that they are in an area without some of the basic materials needed to build it. This can be true of my design, the SunStar. It is built using cardboard, but in some areas of the world cardboard is unavailable or quite expensive. Then you will be faced with designing a cooker that fits the circumstances. By understanding the basic concepts behind cookers, it will easier to build one appropriate to your situation.

The first part of this chapter is a general outline of the building process. Over the years, I have found that I followed a pattern when designing a solar cooker. Look through this outline and see if it can help you in the building process—you might find some useful tips.

After that, I will discuss some of the basic components and materials that I've found useful. This designing information is based on my own experience and is not meant to limit your own creativity. Your creativity and experience are important, especially if you are building solar cookers for areas of limited resources or specific cooking needs.

Many refinements, both practical and aesthetic, can be made.

General Flow Outline

I Define your needs.
   A Where and when are you going to use this cooker?
      1 In your yard, for example, what about trees, shadows, a level spot, etc.?
      2 Is it for camping, backpacking, car camping, fairs, Saturday markets, or home use?
      3 How are you going to transport it?
      4 What time of the year do you want to use it? Is it for only one season (like summer), or is it for any sunny day of the year?
      5 What is the environment like when you want to cook? For example, do rain clouds quickly build up, are there high or gusty winds, or are the weather conditions usually stable?
      6 Do you plan to be around it when cooking or do you want to cook the food while you're gone?
   B What do you plan to cook?
      1 Do you want to use this cooker for all kinds of food?
      2 Is it for cooking one particular way, like baking, pizzas, or cookies?
   C What size cooker do you need?
      1 How many people are you cooking for? Is it going to be more useful to have two medium size cookers, or one larger one?
      2 What kind of cooking pots do you want to use? Will you use more than one pot at a time?
      3 What type of reflector system do you want to use?
      4 A smaller solar box cooker must have have a four square foot window.
      5 For a multi reflector cooker, the size of the window also relates to the final power of the cooker.

Chapter 6: Designing & Building Your Own

A small cooker should have about 100–130 square inches of window. A medium cooker would have 135–180 square inches, while a large cooker might be 180+ square inches.

II Draw the design you've decided on.

   A Define the size of the interior of the cooking oven. If your oven is going to tilt towards the sun, be sure there will be clearance for the cooking pots in the different cooking positions.

   B How are you going to access the food? Will you use a door in back, or use the window for the door? Draw how you will frame the door so you can seal it. Draw how the glass will sit on the box so it does not leak heat or fall off.

   C Draw the necessary framing to fasten the inner and outer ovenbox walls.

   D Draw the kind of rack system you will use. Is your cooker tilting towards the sun? If so, how will you keep the food level?

   E If you're going to tilt the ovenbox, how are you going to hold the ovenbox while it's in various positions? A stand? propping materials? If you're using a stand, does it need wheels, pivots, etc?

   F How are you going to fasten the reflectors to the ovenbox? Are they going to fold down on the ovenbox when it gets put away, or are you going to remove the reflectors for packing away?

   G What shape and angles are you going to use for the reflectors?

III Decide which materials and tools you will need, and gather them.

   A How available are these materials? If you are designing a cooker that other people may want to build, are the materials commonly available?

   B What will you use for insulation? What materials will you use for the inner and outer ovenbox walls, for the reflective surfaces, reflectors? What kind of paint? Does your design use any additional materials, for example: hinges, bolts, nails, etc.?

C What about the cost of the individual and collective parts? Is this an important aspect to you?

D Are the materials you want to use safe when used in the cooker? Are they toxic when heated? Will they cure over time?

IV Building the cooker.

A Ovenbox

1 Frame and cover the inner and outer walls while adding insulation.

2 Place the window and door on the ovenbox.

3 Install the rack system.

4 Paint the inside of the ovenbox.

5 Seal the ovenbox so that moisture does not get into the insulation.

B Assemble the reflectors.

1 Cut them out and apply the reflective surfaces.

2 Attach the reflectors together and attach them to the ovenbox.

C Build a stand if one is desired.

V Test and start using the cooker.

A Set up the cooker and point it towards the sun. Most materials being used will need to cure. They will outgas a small amount of smoke the first few times they heat up. This should clear up soon.

B When the oven is hot, feel the outside for hot spots. Hot spots indicate leaks and the need for more insulation in these areas.

C Check for air leaks around the window or door. Small leaks are not too bad, but large ones should be fixed.

D Start out cooking simple foods. Explore and learn from what you have created.

Before we get into the individual components of a cooker, I want to discuss the building of a prototype. If you have

Chapter 6: Designing & Building Your Own

never built a solar cooker before, building one that meets your needs is a little bit of a hit and miss experience. A prototype will give you a practice run, and a feeling of its power. This will give you important insights into building your final cooker. Cardboard is easy to build with, and is an inexpensive material for a prototype.

The Ovenbox

The ovenbox defines the cooking area. Traditionally, the goal has been to make the inside space as small as possible, while still having the ability to cook the amount of food you want. By doing this, you minimize the size of the ovenbox walls, which minimizes the heat loss. By making it as small as possible, you also make it light weight, and thus it is easier to move it around. When minimizing the size of the ovenbox, remember to make it large enough to fit the pots you want to use. If your cooker tilts, there must be room for the pots on the rack in all cooking positions.

The walls of the oven are far more insulated than the glass. Realize that most of the heat loss will be through the glass, and not through the oven walls. So, make the inside space large enough to be comfortable to use. A little extra room inside won't cause a noticeable loss in cooking power.

There are many possible materials to cover the inner and outer ovenbox walls. These include wood, veneer, plywood, sheet metal, steel or aluminum (a favorite source for aluminum is from newspaper printing presses), and cardboard or fluteboard (a plastic cardboard). The list goes on.

For the inside walls, look for a material that can take the heat. Don't use plastic materials here. Sheet metal or wood works very well on the inside of cookers. Cardboard can also withstand the heat, because the temperature of most cookers is below its 451° F ignition temperature.

For the outside, you might want to look for a material that is resistant to the elements. Painting the outside of the box can be an important part of weather-proofing your cooker. Most of these sheeting materials for the oven walls will need an inner frame work to give you the necessary surfaces to attach the walls. Also the window and the door need to be framed in. The spaces between the inner and outer walls will give you room for the insulation.

Insulation

Light reflected into your oven box strikes the dark surface and turns to heat. To build up enough heat to cook, we must put insulation in the walls. This slows the loss of the heat we need for cooking.

The best known insulation is a vacuum, and in the future we will probably use this, but for now it would be too expensive. The other good insulator is still air. If air moves, it will transfer heat and loose its insulative qualities. This is the reason that most insulations create dead air spaces. Insulations that can be used for solar cookers include: household fiberglass insulation, cardboard, crumpled newspaper, and even dry straw or fluffy cotton.

Don't use foams, like Styrofoam or other insulative foams. These were not made for the heat and will start releasing cyanide gas at temperatures above 180° F. There may be a foamboard insulation designed to handle the heat of an oven. This material could be quite useful. A foamboard would make cookers very easy to put together.

I tell people to avoid foamboards any way, because if other people see you using them, they will likely think all foamboards are alright to use. If you look up the specifications of a certain foamboard, and find it can handle the heat, try using it. Tell people who are interested in it not to use just any foamboard, but specify which type to use.

Aluminum foil, though not really an insulative material, can be effectively used with insulation. Aluminum foil acts as a radiant heat barrier, reflecting infrared heat back into the ovenbox. This helps to increase the power in a solar cooker.

Access to the Food

Your ovenbox must include a way to access the food. There are two ways to do this for most common solar cookers. One is to use the window as the door. The other way is to build a door in back or on the sides. Both of these methods

Chapter 6: Designing & Building Your Own

of access have their good and bad points, so it will be up to you to decide how you want to handle this.

It is easier to build a solar cooker using the glass as the access to the food. You can have the glass removable, or it can be in a frame that is hinged to the ovenbox. Either way, using the glass as the door requires you to take some care around the seal between the ovenbox and the glass to minimize air leaks. One problem associated with using the glass as a door is that you have to work in front of the cooker, so you may be exposed to the sun's glare. Another problem relates to the reflectors. You're going to have to either reach through the reflectors to deal with the food, or remove the reflectors first.

Putting a door in back is a little harder to build, usually including extra framing in the ovenbox walls, and the use of hinges and door locks. The main problem to be aware of is convenience. Design your door to make it easy to use. The design

Above: With the SunStar access is through the window.

will be especially important if the cooker tilts towards the sun. There must be easy access to the food while the cooker is in any position. The benefit of having the door in back is that you don't have to work in the sun's glare. Also, the glass can be sealed to the box. This means fewer problems with air leaks. The glass will also be better protected from accidental breakage.

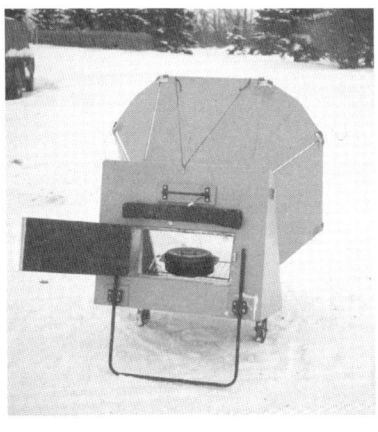

Above: Howard Boldt put his access door in the back.

Glazing

Most solar cookers use some kind of window to let sunlight in. At the same time, the window hinders the escape of the heat necessary for cooking. Glass is by far the most common glazing material.

There are many good reasons for using glass for solar cookers. For most people, it is cheap and readily available at glass stores. Also, with a little practice, glass can easily be cut, so you can use recycled glass window panes. Glass is clear, easily letting sunlight into the ovenbox. Glass can be scrubbed clean without scratching. It does not degrade over time.

Glass does not have a very high insulative value, but it is good enough to make solar cooking possible. Glass can also generally take the heat of cookers. There are usually no problems with breakage from heat with most solar cookers.

Some solar cooking designs can push glass to its thermal limit, where it will start to break. If extra power is something you are after, you may want to use a tempered glass. Tempered glass is much stronger than regular glass, but it is also a lot more expensive. It has to be special ordered from glass companies. Tempered glass can not be easily cut, so it has to be made to size, thus the higher cost. If one was to mass produce a solar cooker, the cost would come down. Then, this could become a very cost effective glazing.

If you want to build a really large cooker, new glass could be one of the most expensive parts. Knowing this, you might want to go to the dump or recycling yard. There, you can buy a large piece of glass, maybe even tempered glass, or double paned (thermal pane) glass. Then, design the cooker around the size of the glass.

For solar cookers using a relatively small piece of glass, it's much better to design the cooker the way you want it. Then, go to a glass store and have them cut a piece, unless you are good at cutting glass yourself. Don't design a solar cooker around a piece of glass you already have, unless it happens to be a good size. I've seen people go to great

Chapter 6: Designing & Building Your Own

lengths to design a cooker around a piece of glass just to save a couple of dollars. They often end up with a cooker that fits the glass but is not useful.

Glass for a medium size cooker is not that expensive. The glass usually costs 2 to 5 dollars. Have the people at the glass store cut a single or double strength piece of glass. I prefer the double strength piece because it is a little stronger.

When glass is freshly cut, the edges will be very sharp. Some glass stores will buff the edges for free, if you ask. If not, just rub a hard rock or piece of metal over each edge. This will make it safe to handle. There are two edges that need to be dulled on each of the four sides of the glass.

Some designers are looking towards ultraviolet resistant, and high temperature rated, plastics as a replacement for glass. An advantage is that plastics are less breakable.

The main reason I don't like plastics is because they scratch so badly, even so-called scratch resistant varieties. In use, solar cookers get dirty with dust and baked-on oils. To clean the glass, it has to be scrubbed. When trying to clean plastics, you create permanent scratches that add up over time, and interfere with light transmission. Also, oils have an affinity for plastics. It's hard to get them clean, so you tend to scrub harder, causing more scratches. Even though I'm attracted to the idea of using plastic glazing I haven't found a practical one for my style of cooker.

Another problem associated with these plastic glazings is that, over time, the sunlight will degrade the plastic. When it wears out, most people will have a much harder time finding new plastic than if they were using glass.

Some people like the idea of using two layers of a 3M film that can take the heat and is UV resistant. This might have some practical use, due to the light weight and non-breakability. I still have some serious questions about how they keep it clean.

Building a Thermal Pane

One of the quickest and easiest ways to gain power in a

solar cooker is to create a thermal pane, often adding 50° F or more to an ovenbox. A thermal pane consists of two pieces of glass with a small air space in between them. The air gap should be 1/4 to 3/8 inches wide. A thermal pane can be created in many ways.

If you have access to woodworking tools, you can build a window frame, and cut two sets of grooves to hold the glass.

Another way is to use silicone. Put one piece of cleaned glass down on a flat surface, then put four spacers of the desired air gap width in the corners. Next, put a bead of silicone around the edges about as thick as the air space will be. Then lower the other cleaned glass onto the silicone and spacers and let dry. When dried, remove the spacers and fill in the gaps around the edges with more silicone.

As a side note, silicone can easily take the heat of most solar cookers, so it can be used to fill any air leak.

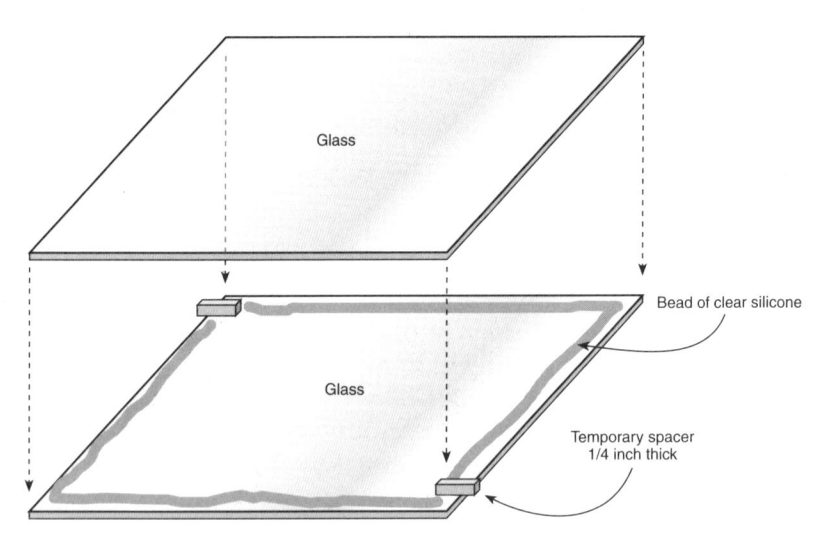

Above: Building a thermal pane.

When you use a thermal pane, you will experience significantly more power. There is a potential that you may push the limits of glass and the inside piece could break. If you want this extra power, you might have to use tempered glass. Personally, I find a single piece of glass works fine and rarely use a thermal pane anymore. If you are in an area with poor sun conditions, or if you want extra power in the winter, then go ahead and use a thermal pane.

Paint

When light hits a black surface it changes to heat. This is the reason for painting the inside of the ovenbox flat black.

Some care must be exercised in choosing a paint. In the USA it is illegal to sell paint that includes lead, but this is not so in some countries, so be aware.

I like painting the inside of my ovens with a black tempera paint. This is kid's finger paint that can be found or ordered through stationery or crafts stores.

Also, most hardware stores sell a paint to be used on barbecue units, which can be used for solar ovens.

Most quick drying, cheap, flat black spray enamels will work, as the solvents will fume off as the cooker is heat cured, leaving just the paint. Most paints will smoke a little at first, until they have been heated up a few times.

Don't paint surfaces that will come in direct contact with food.

I've heard of people, in areas without easy access to paint, using soot mixed with cooking oils instead.

It is not necessary to paint the entire inside of the box black. Sometimes you might want to cover just the bottom black, and cover the inside side walls with foil. Light hitting the side walls should continue downward until it strikes the bottom of the cooking pots. This will increase the power of the oven noticeably. Again, this extra power may push the glass in your cooker to its thermal limit.

The other place to use paint is on the outside of the cooker. This will increase the cooker's resistance to the elements and make the oven look nicer. For this, choose whatever paint you want.

Stands

A stand for your solar cooker can make it easier to use. This is especially true if your cooker weighs too much to move around easily. Also, some cookers are designed to be tilted towards the sun, so a stand can make this easier. This is one area where your own designing skills can become important, because there are so many types of potential stands that can be made. Wheels can be included to make it easier to move the cooker around. A lazy susan, to turn your cooker from east to west, can make focusing more convenient. A cradle system, to create the up and down motion, also helps ease focusing.

Above: Jim Reiman's cooker has an example of a quality stand. A central axis provides for tilt. Casters provide for azimuth.

Racks

All solar cookers need some kind of a rack system to hold the cooking pots level. A rack is used also to keep the cooking vessels off the bottom of the ovenbox, so air can circulate around them.

On cookers like the solar box cooker, it can be as simple as a cake rack.

In cookers that tilt, a little more thought must be put into the rack system. You want to keep your food level and avoid spills. This can be accomplished by securing the rack to the stand and let the ovenbox pivot around the food. You might also secure one side of your rack to the front of the ovenbox, and use a hook and chain to hold the other side of the rack in different positions. You might also try a

Chapter 6: Designing & Building Your Own

system where the rack freely swings from some point in the ovenbox, so that if you keep your food centered on the rack, it will find its own balancing point.

Reflectors

As you've seen in the chapter on cooking designs, there are many possible styles for reflectors. They are all simple variations of geometric forms that reflect extra sunlight into the cooking area. Here are the three common styles used in solar cookers.

The first one is the simplest. It is used in the solar box cooker designs. It consists of just one flat reflector that is hinged to one of the long sides of the ovenbox. This reflector can fold down over the glass top of the ovenbox or can be opened to reflect sunlight into the box. It is held in place by a prop stick. To

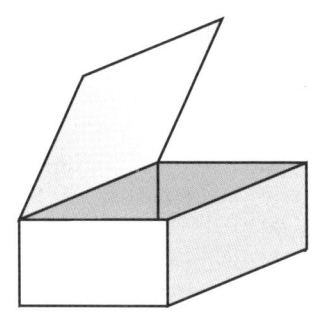

Above: The single reflector solar box cooker.

keep the wind from dislodging the stick, a string is tied from the outside edge of the reflector to the far side of the ovenbox, keeping tension on the prop stick.

The next style of reflector is traditionally used on the multi-reflector style cooker. It consists of four reflectors the same size as the glass. These reflectors are hinged to the ovenbox along each side of the window. They can close down on the window of the ovenbox for packing away and to keep the food warm. These reflectors usually open up

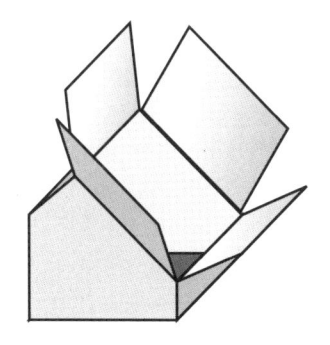

Above: Multiple square reflectors.

to a 60 degree angle, which allows the maximum amount of sunlight to be reflected onto the glass. Sometimes trian-

Heaven's Flame

gle wedges are placed between the four reflectors to add more reflected sunlight to the ovenbox. This creates a slightly higher power.

If you use this style, your main design problem will be stabilizing the reflectors in the cooking position. Bolts, metal brackets, strings, sticks, and props are some of the methods you might use. I used to build my first cookers with this reflector style, but never found a method of stabilization I liked. With nuts and bolts, it just takes too long to set up, which takes away from the enjoyment of using solar cookers. If you design an easy way to stabilize this reflector method, it can be nice because of the way the reflectors fold down onto the ovenbox.

The reflector style I use has four trapezoidal shaped reflectors. By virtue of the shape, these reflectors are much more stable. These reflectors must separate from the ovenbox for packing away. In the next chapter on building the SunStar solar cooker I will go into much more detail on how to build this style.

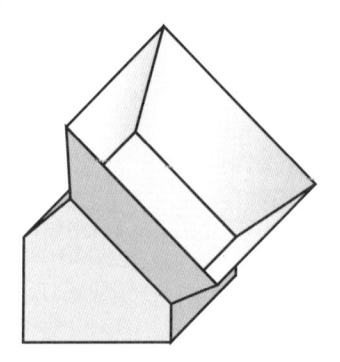

Above: Trapezoidal reflectors.

I found that the solar cooker had more power if the traditional reflector angle was changed from 60 degrees, towards a steeper angle, closer to 67 degrees. The sunlight reflecting off the reflectors strikes the glass at a correspondingly steeper angle and more of the light penetrates the glass instead of bouncing off of it.

Below: Steeper reflector angles transmit more light through glass.

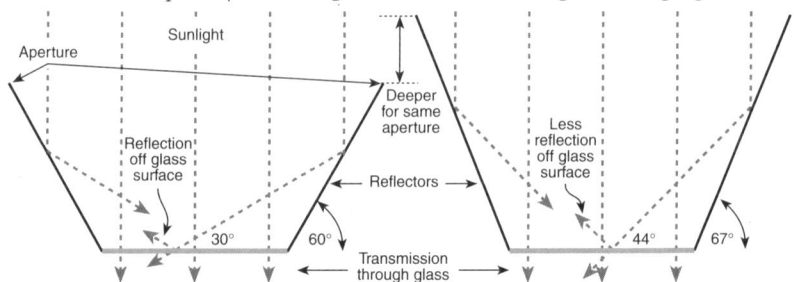

Chapter 6: Designing & Building Your Own

Reflective Materials

I know of four options. Mirrors are probably the most reflective, but in practical use they have serious limitations. They are heavy and breakable, and in most styles, would tend to tip the oven over. One exception is the Solar Chef. This cooker, by its design, makes very good use of mirrors. Thus, it benefits from the high reflectivity and long life of mirrors. Most of the other designs have had to shy away from this material.

Certain polished metals can also be used. These can include aluminum or stainless steel. Stainless steel tends to be harder to work with, heavier, and more expensive than aluminum, but some very nice looking cookers have been made from it. Aluminum can come in many different thicknesses and reflectivity. Sometimes the word "anodized" is associated with aluminum. Anodizing is a process whereby aluminum is treated to make the shine resistant to corrosion. Anodized foils are also commercially available.

The next material is an aluminum polyester laminate. You've probably seen this. It's that shiny plastic film, like the kind they use for space blankets or the inside of some potato chip bags. This is relatively inexpensive compared to mirrors or metals, but it is also harder to find. It will also degrade in the sun over time, unless you buy special UV resistant film. The Hybrid Solar Kitchens use this material in their parabolas, due to its high reflectivity and light weight. They replace it when it wears out. Sometimes craft or stationary stores can order this material for you. From my own experience, its unavailability in common stores, combined with the fact that it wears out, makes this alternative not very attractive. Maybe as cookers grow more common, it will become more commonly available.

The last material commonly used for solar reflectors is aluminum foil. When I first started building and designing solar cookers, I had a strong bias against foil. However, after using it, it has become my favorite reflective material because it works well, it's cheap, it's widely available, and it lasts fairly long.

When light strikes the foil straight on, it is about 67% reflective. When light strikes foil at a shallower angle, as it will in a solar cooker, its reflectivity closely approaches mirror quality. Earlier, I talked about the danger of eye damage due to glare from highly reflective surfaces. This is another reason I like aluminum foil; it seems to give off a much softer glare.

Backing Materials for the Reflectors

When using most of these reflective materials you will have to have some kind of ridged backing for it, the exception being sheet metals that hold their own ridged shape. These backing materials should be light weight if possible. Veneers, thin plywood, cardboard, or sheet metals can work. The new plastic cardboard, fluteboard, might also be ideal because of its light weight and water resistance. Rubber cement or diluted white paper glue or even wheat paste can be used for attaching the reflective films to the different backing materials. I will talk more about attaching foil to cardboard (or wood) in the next chapter.

Chapter 7

Building the SunStar

The SunStar is my addition to the solar cooking world. Like all the designers of solar cookers in this book, this cooker reflects my own life experiences.

During the 80's, I was living a nomadic life, with very few resources. As a designer of solar cookers, poverty was quite useful. I became intimately aware of some issues concerning basic survival. This gave me a better understanding of the lives of some of the people who might find solar cooking very necessary. I was finding that solar cookers were helping me to survive. On sunny days, I could cook inex-

Left:
The SunStar.

pensive, simple, but high-quality meals. On cloudy days, I was cooking more expensive, quick-cooking junk foods, with a small propane cooker. The propane meals were never as nutritious or satisfying. These experiences helped define my design goals.

I wanted to design a solar cooker that:
1 Cost as little as possible;
2 Could be made by almost anyone, without regards to building skills;
3 Required no special tools or a workshop to build;
4 Used commonly available materials;
5 Had useful cooking power for a family, but also power that is impressive, to demonstrate to people know that solar cooking is a real alternative;
6 Would not intimidate people, but instead, by its simplicity, let them know that they too could design and build a cooker to fit their needs;
7 Was fun and convenient to use.

The SunStar fulfills all of these design goals. Over the years, this cooker has become a part of my life. The food it can cook so deliciously has kept my family looking forward to those sunny days, when we can let the bright rays of the sun serve us our meals. At this point, it is not the issue of survival that keeps us cooking in this way. It is definitely the taste of the solar cooked meals keeping us looking to the sun. The only way we can share the unique quality of solar cooked meals is to help you build one also. In this chapter, I will teach you as clearly as possible how to make one on your own.

Searching for the Right Boxes

We will be describing how to build a medium size SunStar solar cooker. You can build smaller cookers with this information, or much larger ones as you wish. Medium sized cookers, in most cases, are the most useful in the family setting. They are able to cook nicely sized family meals. Even if you feel that you need more cooking potential than one medium sized cooker, you will probably find that two medium sized solar cookers are more useful than one larg-

Chapter 7: Building the SunStar

er cooker. Though larger cookers can be useful in special circumstances, they are also a little more awkward to use.

The first step in building the SunStar is to go on a treasure hunt. Look for those cardboard boxes you can magically transform into your own cooker. At first, you will be looking for two specific boxes which create the ovenbox. One box ends up being the inner ovenbox, and the other becomes the outer ovenbox. The source of these boxes may be behind any store. Larger stores often recycle their boxes, so you might ask to look in their storage room.

One of the variables in this design is whether you want a rectangular ovenbox (these boxes are easier to find), or a square ovenbox (these boxes are harder to find, but the reflectors fold up more easily). Both shapes are useful; maybe you could build one of each. The smaller, inner ovenbox defines the cooking area. For a medium sized ovenbox, when you multiply the length times the width (width should be at least 9"), the area should equal 120–180 square inches.

For example, a good inner box might have a width of 10" and a length of 14", so 10 x 14 = 140 square inches. Another example is a box that is 12" wide and 12" long, so 12 x 12 = 144 square inches.

Another requirement for a good inner box is that it should be at least 8", better yet 9–12", deep.

The outer ovenbox should be 2-3" larger than the inner box in all directions: length, width, and depth. For example, if the inner box is 10" x 14" x 10" (L x W x D), then the outer box could be 12" x 16" x 12" (or slightly larger).

If you can't find the right boxes, you can easily alter the size with some cutting and gluing (see page 108). Your cooker will be stronger if you find the right size of boxes, without this cutting and gluing, so try your best to find good sized boxes.

fig. 1

After finding these two boxes that define the ovenbox of the cooker, gather

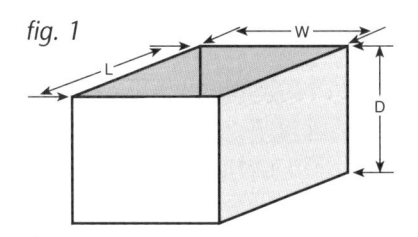

5–7 more medium sized cardboard boxes that can be cut up for insulation.

To make the reflectors, you will need to find four flat pieces that are about 2' x 3' wide. These pieces are cut from large boxes, such as washing machine or bicycle boxes. It is best that this cardboard is single strength rather than double strength. Double strength cardboard is two layers thick and thus is a lot harder to bend and work with.

Other Materials

Next, gather the other materials you will need.

1. About 8 oz. of white paper glue (like Elmers). Homemade glue can made by mixing 1/2 cup of flour, 1 tablespoon of sugar into one cup of water. This mix must be gently heated while stirring until it thickens. When dried this glue has been shown to be inedible to insects, at least in Arizona, and is as strong (or stronger) than Elmer's white glue.

2. One small roll of 18" wide aluminum foil, preferably the heavy duty kitchen foil found at grocery stores.

3. A small amount of flat black paint (see "Paint" in previous chapter).

4. One piece of double strength glass, 1/2" larger than the length and width of the inner ovenbox, so it can sit on the top rim of this box and completely cover it, thus enclosing the cooking area.

5. A baking tin that will fit inside the ovenbox to serve as a rack.

6. Some thick string, maybe a pair of shoelaces.

7. Some strong cotton cloth from recycled clothes.

8. (For a square cooker only) About one yard of elastic. This can be found at sewing stores. It comes as a stretchy ribbon of various widths. Buy the one that is about 3/8" wide.

Chapter 7: Building the SunStar

Putting the Ovenbox Together

1. First, work with the larger, outer ovenbox. Cut pieces of cardboard (from the extra boxes you collected) and place them in the bottom of the outer ovenbox until the cardboard layers are one inch thick (see figure 2). This will take about seven layers of cardboard. Some layers can be made from pieces of cardboard; they don't all have to be the full size. Place one or two layers of aluminum foil among these cardboard layers to act as heat reflectors (optional).

2. Next, position the outer box's upper flaps. Bend two opposite flaps inward and down. Bend the other two flaps outwards (see figure 3). If you are building a rectangular cooker, bend the two longer flaps outwards.

3. Now position the flaps of the smaller, inner ovenbox. Fold all four of its upper flaps all the way out and down. Place this box inside the outer box (see figure 4).

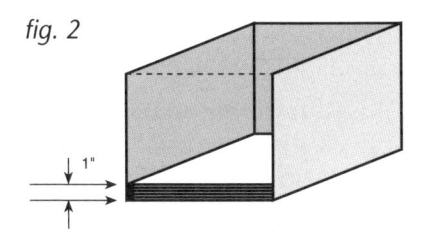

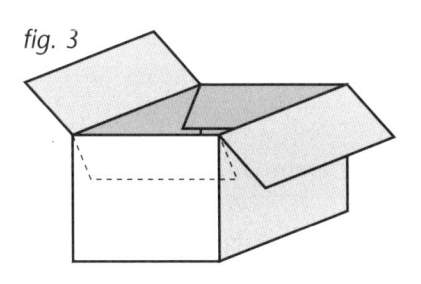

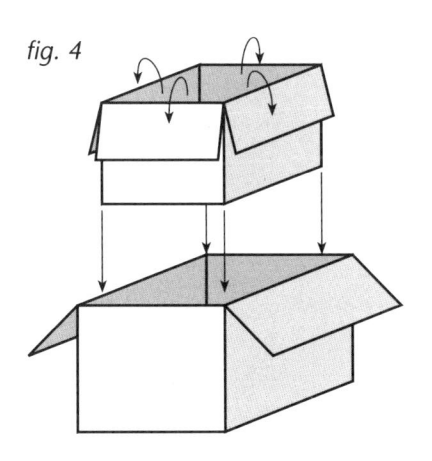

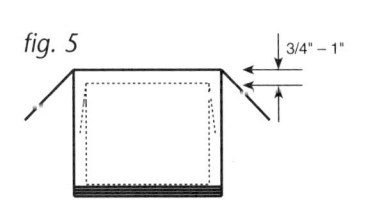

fig. 2

fig. 3

fig. 4

fig. 5

91

Heaven's Flame

The upper flaps must end up between the two boxes (see figure 5).

4 The top rim of the inner box should be 3/4"–1" lower than the top rim of the outer box (see figure 5). If it isn't, adjust the number of layers of cardboard in the bottom of the outer box.

5 The next step is to cut other pieces of cardboard (from the extra boxes) and stuff them between the inner and outer ovenbox. These cardboard filler pieces are the insulation for the ovenbox. Designate one of the four side walls to be the upper side wall and one the lower. The upper sidewall is defined as the higher sidewall when the ovenbox is tilted towards the sun. Designate which are the upper and lower side walls of the outer box from the sides with the flaps folded out. When you start placing the filler pieces into the side walls there are a few things you should keep in mind:

   A All four side walls should end up with similar thicknesses, about one inch.

   B It looks nicer if most of these filler pieces are doubled over, so that the cut edges of cardboard don't show (see figure 6). This is purely an aesthetic step; the filler pieces don't have to be doubled over to work. To bend cardboard where you want, crease a line with a blunt point and bend in on the crease (see figure 7).

   fig. 6

   fig. 7 — Crease a line

   Fold on crease

   C The tops of the filler pieces on the lower side wall should be arranged so that

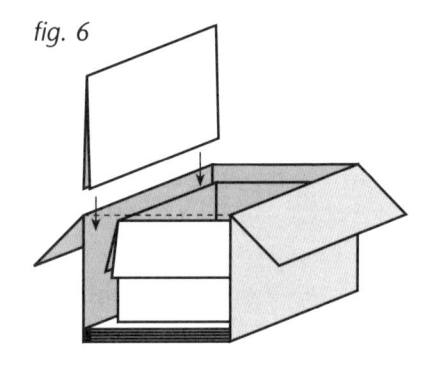

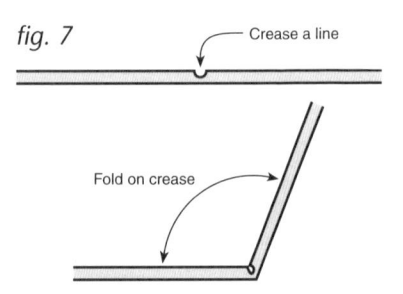

Chapter 7: Building the SunStar

when the glass is placed on top of the inner box, it will still cover the inner box completely. When the ovenbox, with the glass in place, is tilted towards the sun, the glass will slide down and rest against the cardboard filler pieces in the lower sidewall (see figure 8). When the glass is in this tilted position, it must still completely cover the inner ovenbox in order to trap the hot air.

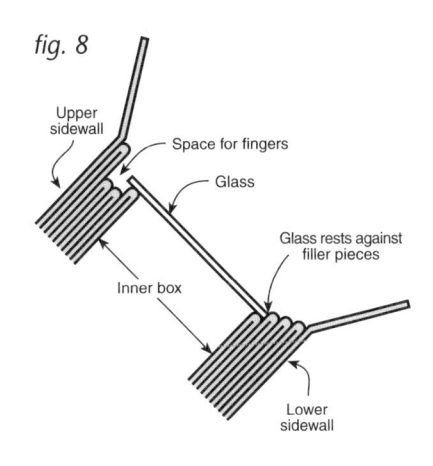

fig. 8

Upper sidewall
Space for fingers
Glass
Inner box
Glass rests against filler pieces
Lower sidewall

D The filler pieces in the other three side walls should be arranged so the pieces next to the inner box are slightly lower than the top rim of the inner box. This is to allow the glass to rest on the top rim of the inner box with no interference from the filler pieces (see figure 8).

E The filler pieces near the outer box should be slightly higher than the top rim of the inner box (see figure 8).

F When the glass is placed on the top rim of the inner box, you should be able to slip a finger under one side to remove the glass.

G Place one or two layers of aluminum foil into the four side walls to act as heat reflectors (optional).

H Continue putting these filler pieces into the side walls until the inner box is held tightly in place. Later, after the oven box is used for a while, the inner box will shrink slightly and will loosen up. The side walls should be repacked at this point, and at any time it becomes loose again.

6 Paint the inner ovenbox flat black (see figure 9). Because the bottom of the inner box heats up the most, it

Heaven's Flame

will show a little wear in a year or so. To protect this surface you can cut another piece of cardboard to fit in the bottom. Paint it black (optional). When this added piece wears out, you can just replace it with another piece.

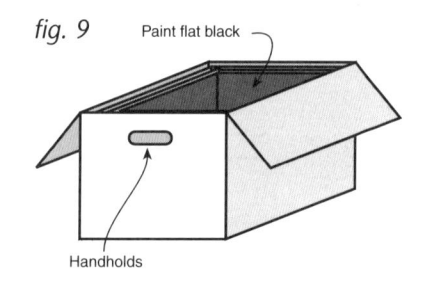

fig. 9 Paint flat black

Handholds

7 Cut handholds out of the outer ovenbox so that the oven will be easier to carry (optional, see figure 9). Just cut through one layer of cardboard and squirt some glue up under the cut to keep it from tearing out.

Building the Reflectors

1 Draw the reflector dimensions onto the four flat pieces of cardboard (see figure 10).

   A The reflectors are based on the size of the glass. A square ovenbox will have four reflectors of equal size. A rectangular cooker will have two sets of equal size reflectors based on the length and width of the glass.

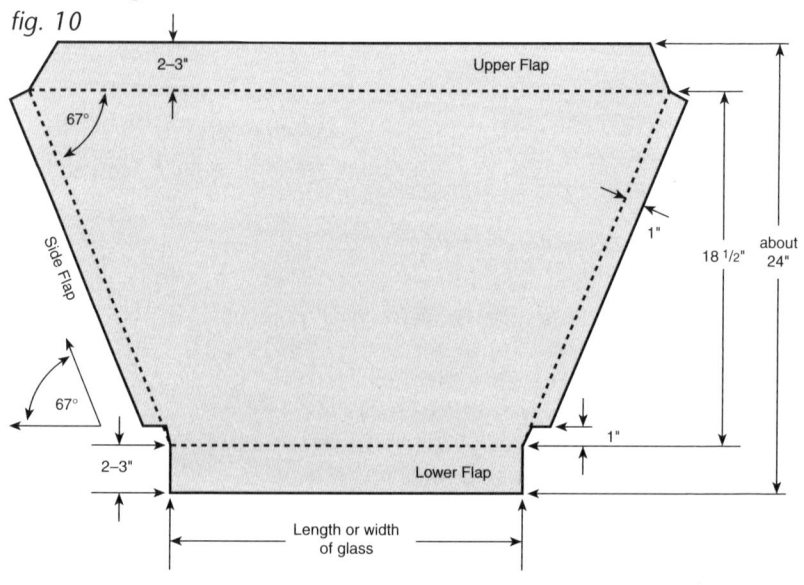

fig. 10

2–3" Upper Flap

67°

Side Flap

1"

18 1/2" about 24"

67°

2–3" Lower Flap

1"

Length or width of glass

Chapter 7: Building the SunStar

fig. 11

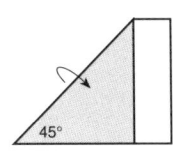

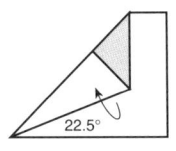

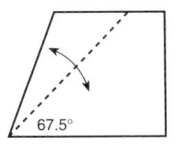

22.5° + 45° = 67.5°

B   The SunStar's reflectors are based on a 67 degree angle. To find this angle, use a protractor. You can also find this angle by folding a piece of paper a couple of times (see figure 11). Take a square corner (90 degrees) of a piece of paper and fold it in half so you have a 45 degree angle. Now fold one of the 45's in half, and you have two 22.5 degree wedges. Unfold the paper and add one of the 22.5 degrees to the 45 degree angle (22.5+45=67.5). Now you will have an angle close enough to 67 degrees.

C   To keep the reflectors fairly durable, the design includes flaps that can be folded over and glued on to the top and bottom edges of each reflector. These flaps should be about two to three inches wide.

D   To connect the reflectors together, you need to leave one inch flaps on both sides of each reflector. They should start one inch from the bottom side.

E   The distance between the upper and lower sides, not including the flaps, should be a little more than 18", say 18.5". This is based on the common size of a roll of aluminum foil.

2   .Once you have the dimensions drawn, cut out the reflectors. Use a razor blade knife or a mat knife.

3   Next, crease a line between the upper flap and the rest of the reflector, and bend in on this crease. Follow the same step for the lower flap and the side flaps. They should all bend towards the same direction (see figure 12).

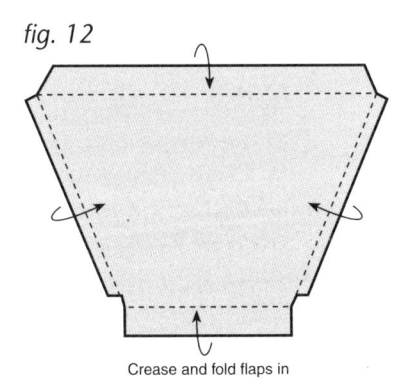

fig. 12

Crease and fold flaps in

95

Heaven's Flame

4. Bend the upper and lower flaps all the way back and glue them to the back side of the reflector (see figure 13). When you glue these flaps, press them with weights until they are dried. Now, you have four separate reflectors.

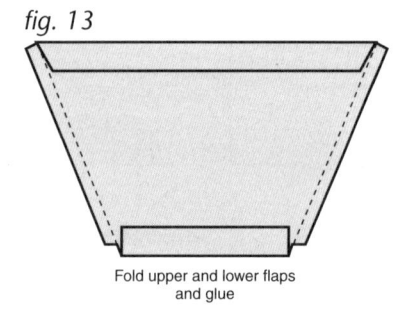

fig. 13

Fold upper and lower flaps and glue

Gluing the Foil on the Cardboard Reflectors

1. For this step, you need a flat working surface. Do a careful job because it will make your cooker look nicer and work better. Lay a reflector on the work surface with the upper and lower flaps down. This allows the clear side, to which you are going to apply the foil, to be facing upwards. Take the roll of foil out of its box and unroll it so that it completely covers the reflector. Rub your finger over the creases in the side flaps. This slightly creases the foil and will show you where to cut it. Cut the foil so that it does not quite reach the side flaps, leaving about 1/4" clearance (see figure 14). This will make it easier to center the foil onto the reflector when applying. The foil must not end up covering the side flaps.

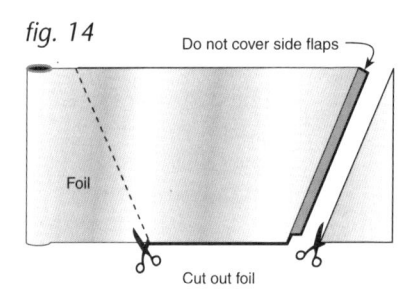

fig. 14

Do not cover side flaps

Foil

Cut out foil

2. Remove the cardboard reflector from the working surface. Lay the cut piece of foil on the working surface, dull side up. Next, make up some of the glue mixture you will use, either the flour paste or the white paper glue. If you are using a white paper glue like Elmer's, take about 5 tablespoons of glue and add 10 tablespoons of water and stir it up. Now dribble 2-3 tablespoons of this glue on to the dull side of the foil (see figure 15). Take a small piece of clean cloth and smear this glue mix over the whole surface of the foil. Be care-

Chapter 7: Building the SunStar

ful around the edges, so that the glue doesn't get on the shiny side. It should smear easily. If not, add more water to your mix.

If you are using the homemade flour paste, apply a thin layer to the dull side of the foil. Again, if it is not easy to spread, add a little more water.

fig. 15
Dot 2 – 3 tablespoons of glue on dull side of foil

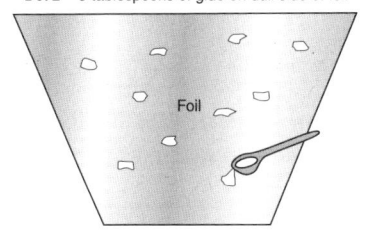

fig. 16
Line up bottom edge of reflector and foil then lower reflector onto foil

3   Next, line up the lower side of the cardboard reflector with the lower side of the foil. Gently lower the reflector onto the foil (see figure 16). Press lightly and turn it over, the foil should stick to the cardboard. If it is not centered very well, gently try to center it with a little pressure from your fingers. If this does not work, peel the foil off at this point. Let the glue dry, put some more on, and try again.

4   Press on the edges of the foil with your finger and pull out any large wrinkles. Small wrinkles are alright. Take another piece of clean cloth and rub the foil onto the reflector from the center outwards (see figure 17). Again, be careful around the edges, so that the glue doesn't get on the shiny side. If some does, let it dry and then rub it off. If a small air bubble is created that you can't rub out, take a pin and prick it and then rub again. Later, if any corner is not glued down well, take some more of the glue and stick it down.

fig. 17   Rub outward on foil side

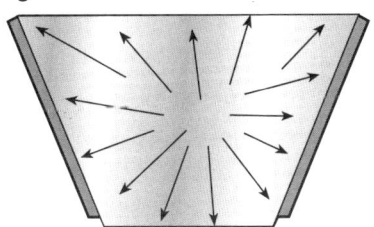

Heaven's Flame

Judge your work now. Does it look nice? If not, peel it off before the glue dries, and try again. Remember, it doesn't have to look perfect.

5 Repeat the process until all four reflectors are done.

Connecting the Reflectors Together

For a rectangular cooker:

1 Arrange the reflectors as they will fit on the ovenbox and set them out on the ground (see figure 18).

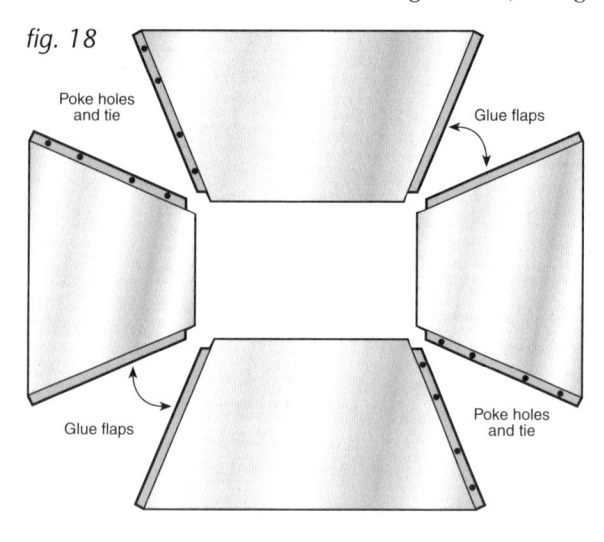

fig. 18

2 Glue two diagonal sets of corner flaps together. You then end up with two sets of two reflectors. When you glue these flaps together, press them with weights until dried.

3 Connect the side flaps that are not glued together. In each of the remaining flaps, poke four sets of corresponding holes. These two corners are tied together with a thick cotton string. Punch the holes near the bend between the reflector and the side flap (see figure 18).

For a Square Cooker:

1 Glue two sets of side flaps together so that you end up with two sets of two reflectors (see figure 19). Press the glued side flaps with weights until they are dry.

Chapter 7: Building the SunStar

2. Lay the reflectors out on the ground as they will be arranged on the ovenbox.

3. Cut off one of the side flaps from each set of reflectors in only one of the corners (see figure 20).

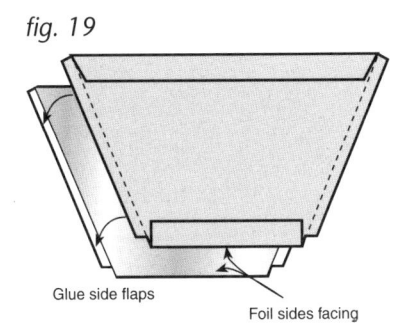

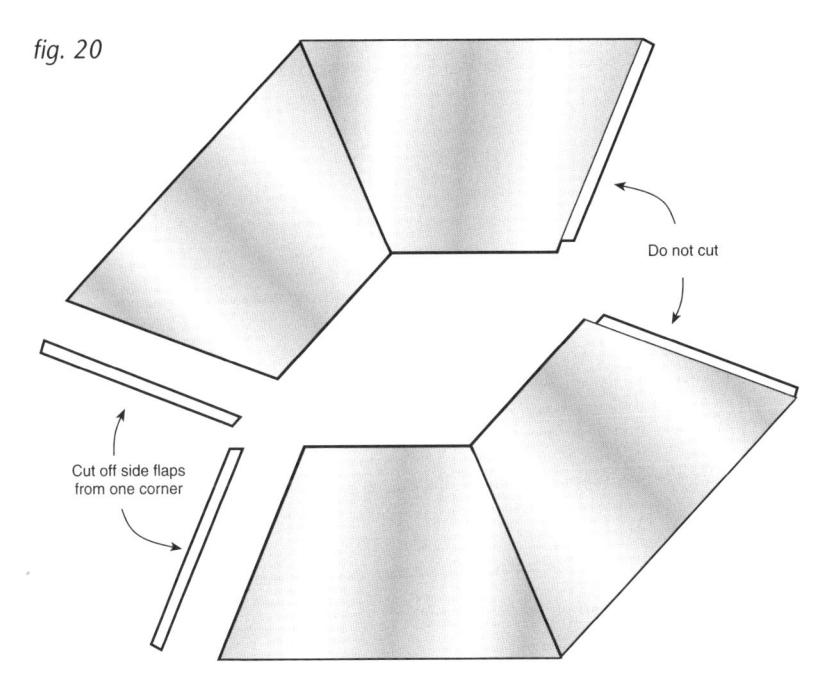

fig. 19

Glue side flaps

Foil sides facing

fig. 20

Cut off side flaps from one corner

Do not cut

4. Arrange the reflectors so that the two reflectors with the side flaps cut off are lined up with their reflective sides down. Cut a piece of cotton cloth, 4" by 18". You will use this cloth to glue this corner together. Apply glue to the cloth and then lay the glued surface down onto the cardboard side of the seam between the reflectors (see figure 21). This will make a cloth hinge in one corner of your reflector assembly.

fig. 21

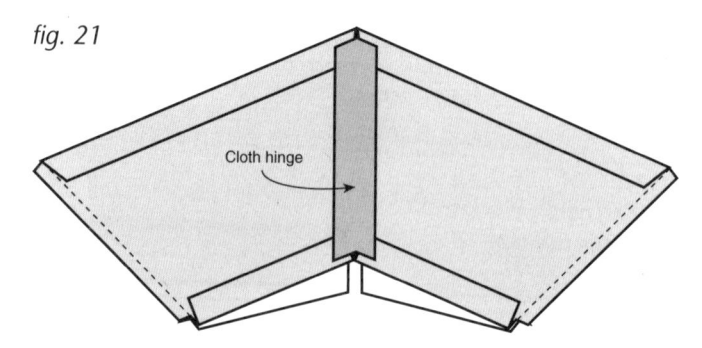

fig. 22

5. When the glue on the cloth hinge has dried, arrange the reflectors so that they are all together in one pile. The cloth hinge should be folded in on itself in the middle (see figure 22), and the two side flaps that are not connected should be lined up on the outside. Poke four sets of corresponding holes in the side flaps that still need to be connected. Poke the holes near the crease. Take the elastic material you got at the sewing store and tie these flaps tightly together (see figure 22). It probably would be easier if you tie two sets of holes near the lower side together and the two sets near the upper side of the reflectors together. Unfold the reflectors and see what you have created.

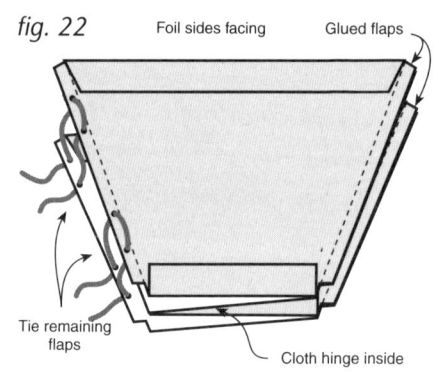

When I wrote the last version of this book, the reflectors were attached to the ovenbox by poking two sets of corresponding holes in both the upper flap of the ovenbox and in the upper reflector. Then I used a string and tied the reflector to the top flap of the ovenbox. This works fine. I have also found that I could create a slip-in piece to easily attach the reflectors to the ovenbox. By inserting the slip-in piece between the insulating layers of cardboard in the upper side wall, the reflectors can be secured to the ovenbox. I still tie the reflectors to the upper flap of the ovenbox, but only when it's windy.

Chapter 7: Building the SunStar

Creating the Slip-in Piece

1. Cut a piece of cardboard 16" wide and the same length as the glass. For example, if the glass is 14" long, then this cardboard piece should be cut 16" by 14".

2. Crease two lines, on the same side of the cardboard, to split the 16" into three sections with widths of 6", 6", and 4" (see figure 23). Bend in on these creases.

3. Cut a piece of cotton cloth 6" wide and the same length as the glass. Put glue on 3 inches of this cloth and lay this glued cloth onto the first 6" section of cardboard. This leaves 3" of the cloth without glue not on the cardboard. Next, put a little glue down each of the two creases in the cardboard and again on the 3" of cloth. Smear the glue around a little. Fold the 4" section over onto the adjoining 6" section (see figure 24). Then fold these two sections over onto the 6" section with the cloth leaving a 3" strip of cloth exposed (see figure 25). Press the slip-in piece with weights until it is dry.

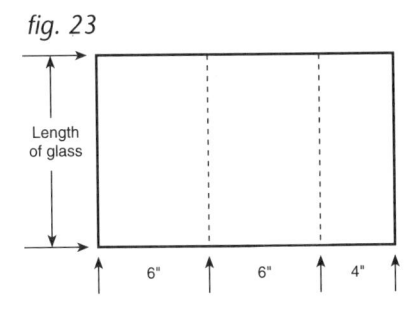

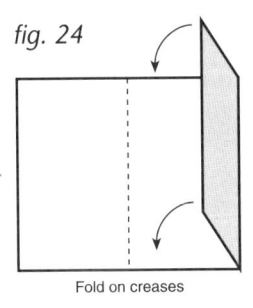

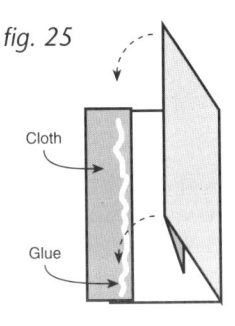

fig. 23

Length of glass

6"   6"   4"

fig. 24

Fold on creases

fig. 25

Cloth

Glue

When you use your cooker you will understand why I fold the slip-in piece this way. You will be pushing down on the slip-in piece to insert it between the insulating layers of cardboard in the upper side wall of the ovenbox. The slip-in piece will last longer if you are pushing down on a doubled over piece of cardboard.

Attaching the Slip-in Piece to the Reflectors.

1. For the rectangular cooker, just glue the 3" width of cloth, sticking out of the slip-in piece, to the lower side of the upper reflector (see figure 26).

fig. 26

2. For the square ovenbox, glue the 3" width of cloth, sticking out of the slip-in piece, to the lower side of the reflector to the right of the reflector's cloth hinge (see figure 27).

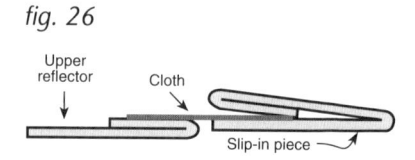

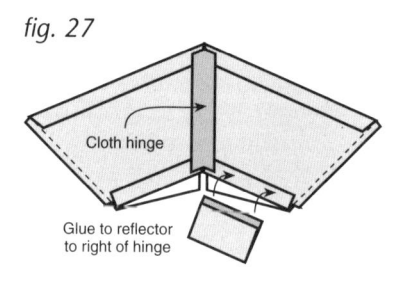

fig. 27

Finishing and Setting up your Solar Cooker.

Push the slip-in piece, with reflectors attached, between the insulating cardboard spacers on the upper side of the cooker. The reflectors should hang in place off of the cloth hinge attached to the slip-in piece (see figure 28).

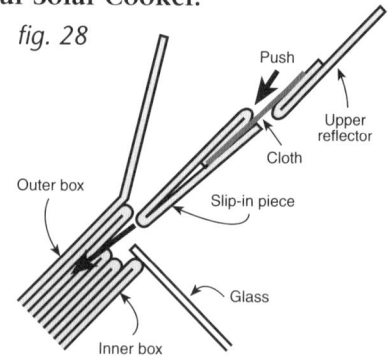

fig. 28

The Rack

A dark baking tin is used as a rack to hold the cooking vessels and to catch any boil over. The rack is propped level against the bottom and the lower sidewall of the ovenbox.

It's nice if the rack puts a slight pressure on the oven's left and right sidewalls to add stability. If the baking tin has handholds, they can be bent downwards to make it fit better.

For example, let's assume that the inner ovenbox is 14 inches long. A baking pan that is slightly larger than 14 inches, when you include the hand holds, would be a perfect size. You can slightly bend the handholds down until the pan is just barely more than 14 inches long and then slip the pan into the ovenbox. It will now put a slight pressure on the sidewalls for stability.

Any baking tin that fits in the cooker can work. If the pan does not put a little pressure on the oven's walls, you will have to be a little more careful in arranging the cooking vessels so that they don't spill. You might have to set-up a propping system inside the ovenbox at some cooking angles.

You can build your own rack out of wood or bent metal. Paint the rack black.

When I build a cooker now, I start with the size of a traditional baking tin and find the necessary boxes for the oven to fit the pan. At first, though, it's more important you find any boxes that will work to get you started. In the future, you might keep your eyes open to find boxes that will fit a particular baking tin that you'd like to use.

The Glass

Place the glass on the top rim of the inner ovenbox. If any of the cardboard filler pieces are interfering with the seal between the glass and the rim of the inner box, correct them. Next, lightly push down on the four corners of the glass when it is in place on the top rim of the inner box. If the glass rocks, it means two of the corners are high. Press down on the high corners until the glass no longer rocks.

Take a close look at the fit between the glass and the cardboard rim. Sometimes there is a high or low spot. Correct this by pushing down on higher spots on the top rim of the inner cardboard box. Just try to get as good a fit as can be easily achieved. Over time, as the cooker is used, steam will naturally make the glass seal better.

It should be easy to slip a finger under one side of the glass when it is lying on the ovenbox. This will allow you to easily remove it to access the food. You may have to cut some of the filler pieces to make it easier to slip your finger under one corner.

When the glass is hot, you should still be able to still handle it, if you only hold it by the edges. If it is still too hot for you, use a cloth to protect your hands. Some people tape the edges to make the hot glass easier to handle. You can do this, but if you learn to handle the edges only, with the palms of your hands, the tape will not be necessary.

Propping the Cooker

Use rocks or other objects (blocks of wood, shoes, rolled-up clothes, etc.) to tilt your cooker towards the sun. You must prop it both in front and in the back (see figure 29). If you only prop up the back, wind can easily tip it over and make an unnecessary mess in your cooker.

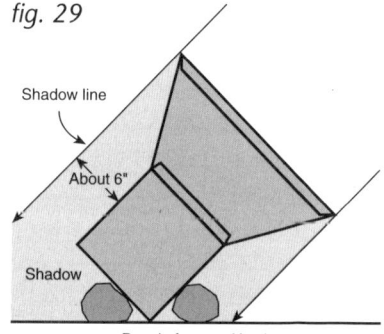

fig. 29

Shadow line
About 6"
Shadow
Prop in front and back

Aiming the Cooker

Never look at the sun to aim your cooker! Always use the shadows created by the cooker. For the east-west motion of the sun, use the shadows on the sides of the cooker.

For the vertical motion of the sun, stand behind the cooker and touch the ovenbox. Watch the shadow line that crosses your arm. A focused cooker will have the shadow line cross your arm about 6" from the ovenbox (see figure 29). By watching how the shadows move over time, you will learn how to aim the cooker ahead in the path of the sun.

Wind

For windy days, poke two corresponding sets of holes in the flaps of the ovenbox and in the upper and lower reflectors. Tie each of the reflectors to a flap with string (see figure 30). Shoe strings work well. Also don't forget to prop

Chapter 7: Building the SunStar

your cooker securely, in front as well as in back.

fig. 30

Tie collectors to oven box when windy

Curing the oven

The first couple of times you use this cooker it will smoke slightly, but after a few times this will stop.

Inner Box Shrinkage

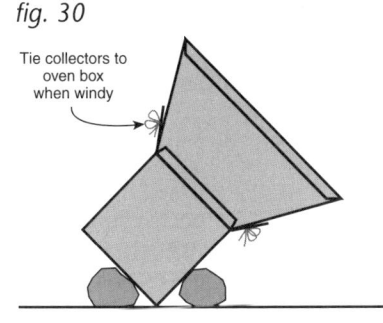

After cooking a few meals, the inner box will naturally shrink a bit. The reflectors' slip-in piece will no longer fit tight enough to hold the reflectors to the ovenbox. Cut some more filler pieces and repack the walls of the oven. Usually after another month or so, the ovenbox will again have to be repacked with more cardboard pieces, but this will probably be the last time.

Important Habit

Your cooker is essentially made from paper, so you don't want to leave the cooker out in the rain or out at night in the heavy dew. You can close the reflectors down onto the cooker after the food is cooked, and carry the cooker into the kitchen. This keeps the food hot and in the kitchen and the cooker out of the weather.

Another Good Habit

Reflectors, when new, will bow due to changes in humidity. Don't worry about this. After a while, when the reflectors get older, they will lose this ability to bow. Also, when storing your ovenbox for extended periods of time, set the ovenbox on top of the reflectors. The weight will keep the reflectors flat.

Above: Folded down, the reflectors insulate the food until ready to eat.

Rebuilding

After a year or so, these cookers might need some maintenance. The most important will be to put a new layer of foil on your reflectors. Usually you can just peel off the old foil and reapply new foil. Sometimes it's hard to remove. A little sandpaper will help.

If the outside ovenbox is damaged, you can slip the ovenbox into a slightly larger box, repacking it with filler cardboard pieces as necessary.

If the inner box is damaged, you can just put another piece of cardboard in the bottom and paint it black.

Also, you can use tape and glue for many simple repairs. Cloth with glue or wheat paste can be used instead of tape for repairs. Weak points can also be sewn with string for added strength.

Cooking Utensils

The SunStar was designed to use recycled jars with lids as cooking pots. They are a little unconventional, but they are excellent solar cooking vessels. Jars are, by far, my favorite pots for solar cooking. They allow you to cook three, maybe even four dishes in an area that could only fit one traditional pot. Jars are cheap, easy to find, and being glass, they are very clean to cook in.

To create good solar cooking jars, paint the outsides black. To be able to see the food cooking inside, put a strip of tape down one side before painting, then peel the tape off afterwards. This leaves a clear strip on the cooking jar.

Sometimes the lids of jars have a rubber seal in them. It is probably best to remove this seal. Sometimes the seal can be scratched off, but it's usually easier if the lids are heated up slightly. You can also burn the seal off.

If you decide to leave the seal on the lid, don't tighten the lid down firmly, so steam can easily escape. If the food is actively boiling, be careful and open the lid slowly.

Some people like to poke a hole in the top of the lid to be sure that no pressure develops. I want you to be careful using jars for cooking, but with just a little care they will be the best pots for solar cooking. I've never had any trouble.

Chapter 7: Building the SunStar

Right: Jars (with lids removed) as cookpots.

The lids will rust over time, so treat them like cast iron pots, by curing them with oil and heat.

To get started, just find a few jars with lids. Wide mouth jars are easier to clean, but because solar cooked food doesn't stick or scorch, even jars with small lids are easy to clean. The jars can be of various sizes from 12–64 fluid ounces. One gallon jars may crack because their walls are too thin for their size. Try to find at least one, maybe two, wide-mouthed 1/2 gallon jars. This is a very useful size. Sometimes you can buy this size jar for a couple of dollars.

When you use jars, there are a few things to consider. Food expands as it cooks, so don't overfill the jars. Also, as the food cools, sometimes the jar will pull a vacuum and make it hard to open. To release the vacuum, take a spoon and pry up under the lid. Sometimes, food particles will cause the lid to stick. If this is the problem, just tap around the top of the lid. If you oil the lid this might help keep it from sticking.

One nice thing about jars is that if your SunStar cooker tips over, only some of the liquid will spill out, not the solid food. This makes it at least a cleaner mess. When your food spills, pour as much of the liquid you can out of the oven-box and set the cooker up again. Dry it out as soon as possible. Take this as a lesson, and figure out how to avoid your next spill.

For casseroles, it will probably be useful to have one casserole dish, with lid, that fits in your cooker.

For baking, use traditional baking pans. Muffin pans also work fine. Bread pans have a limitation in that you have to get heat to the bottom to cook bread well. Tin cans, like 1 or 2 pound coffee cans will cook excellent solar bread. Read more about bread in the next chapter.

Altering the Size of Cardboard Boxes

If you cannot find the right boxes for the inner or outer ovenbox, remember that cardboard boxes can easily be altered. Here are a few examples:

1. Correct length and width, but too deep (see figure 31): Measure desired depth up from bottom. Draw a line around the box at this desired depth. Use a blunt point and crease along this line. Cut corners down to this line and bend sides out. Cut the excess off.

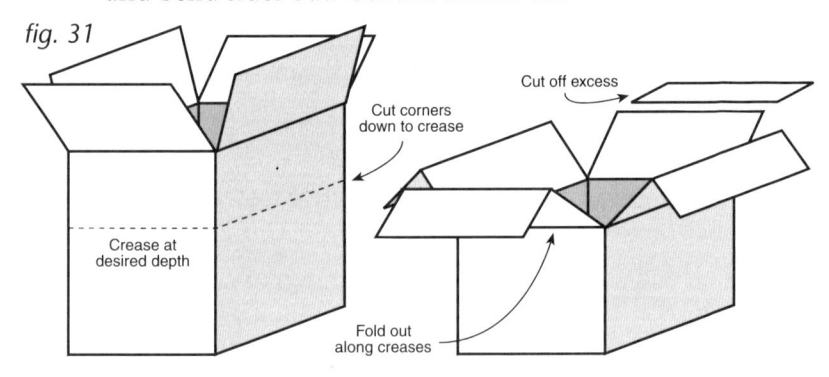

fig. 31

2. Correct length and width, but too shallow (see figures 32 & 33): Measure desired depth up from the bottom of the box to somewhere on the upper flaps and draw a line. Cut the cardboard off above this line. Turn the box over, break the seal, and fold the flaps out and down.

fig. 32

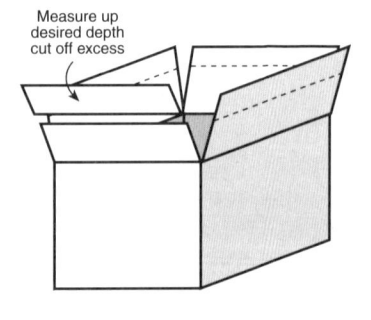

Chapter 7: Building the SunStar

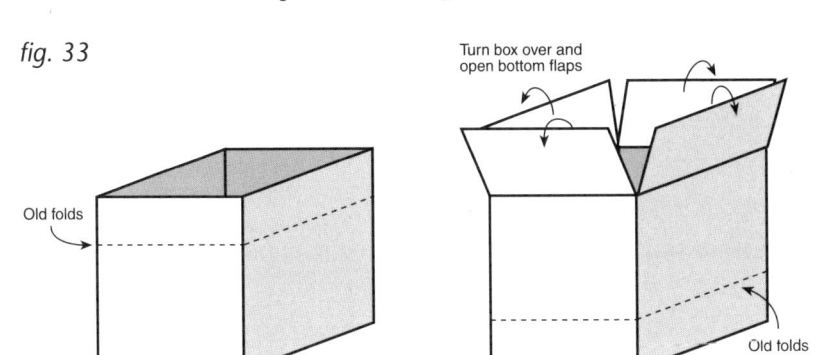

fig. 33

Old folds

Turn box over and open bottom flaps

Old folds

Insert this box into the outer box. To keep the cut flaps in place at the bottom of the inner box, cut another piece of cardboard that will fit snugly into the bottom.

3 Correct depth and width but too long (see figure 34): Cut box at the desired length. Take the short end, and cut off the bottom and the two upper side flaps. Also,

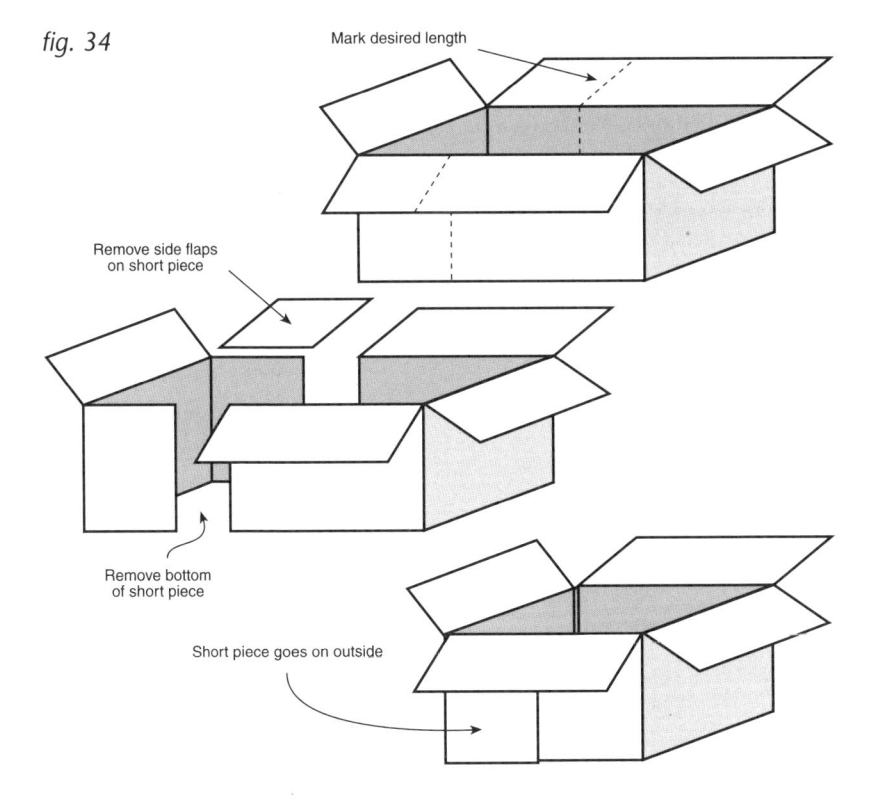

fig. 34

Mark desired length

Remove side flaps on short piece

Remove bottom of short piece

Short piece goes on outside

cut a little off the side walls. Now, put the box back together. The sides of the short end will enclose the long end of the box. Make sure that the top rim of the box is at the same level. Glue in place.

To Increase Power

Other than making a larger cooker, there are two ways to increase the power of a SunStar cooker.

1. Use a thermal pane consisting of two pieces of glass separated by about 3/8" of air. In the last chapter, I told you how to build one. You can expect about a 50° F rise in temperature.

2. On the inner side walls of the ovenbox, glue a layer of aluminum foil. This will add 10–15% more power. The foil on the sidewalls reflects the light onto the cooking pots.

These increases in power may also increase the potential of glass breakage due to heat. If you decide to try these methods to increase the power, and the the glass does breaks, it will break in a squiggly line. This usually happens when the cooker is heated with no food in it, or when a cold wind blows across the hot glass.

Usually, I don't use either of these methods on my personal solar cookers. They are usually hot enough without them, and I have never broken the glass. But there are times these methods of increasing power could be useful and necessary.

Chapter 8

Using the SunStar

Now you have built your SunStar, and it has been cured. You know how to prop the cooker and aim it at the sun. Your cooking vessels are ready. It's time to start learning how to use your solar cooker.

It is not important that you immediately start cooking all of your food with the solar cooker every sunny day. It is much more important that you have fun with the cooker, learn from it, experiment with it, and grow with it at your own pace. What I am sharing is not a fad tool, but one that I hope will become a part of your life. Those of us who use solar cookers today, know that it took years before cooking with the sun truly became second nature. That's why I need to say again: keep it light and fun, so you will continue to use it. One day, solar cooking will become second nature to you also.

For some of you, solar cooking will be new. After you learn to use your cooker, you will find that it is actually a much easier and less time consuming way to cook than traditional methods. For someone just learning to use their solar cooker, this probably won't seem the case. Learning to use it takes effort. You must make a personal commitment to bring your cooker out on sunny days.

I like to tell people to start using their cooker during an extended sunny period, when they will have the easiest time cooking successful solar meals. When it looks like it will be sunny, bring your cooker out in the morning and begin setting food in it. Start simply. Allow solar cooking to gradually grow on you. Develop rituals that will allow

Heaven's Flame

solar cooking to become a part of your life. In the morning, step outside and greet the day. Look at the sky. Is it clear? Are there clouds? Ask yourself what you think you could cook today. Think about the day and what do you have to get done, and imagine how using the solar cooker could fit in to the pattern of the day.

You will have to experiment for a while. You may make some mistakes, usually as undercooked food. Just accept this as part of the natural learning process. You will gain confidence and experience as you continue to use the solar cooker.

Let's pretend that sunny day has arrived, and you are ready for your first solar meal. Maybe it is a day off, and you have time to do some gardening. You've stepped outside and seen the clear blue sky with the sun rising brightly. With some natural anticipation, you bring your cooker out.

When you look for your cooking spot, imagine the arc of the sun and think about shadows that will be created by trees. You may have to move your cooker sometime during the day to keep it in the sun.

Put something easy in your solar cooker, like a quart jar, or better yet a half gallon jar if you have one, with a cup of beans and 2 1/2 cups of water in it. Point the cooker towards where the sun will be in an hour or so. Use the shadows created by the cooker to aim it. When the oven-box is propped in front and back, go about your normal morning ritual. In an hour or so, come back and check it out. Look again at the shadows created by your solar cooker, see how they have moved while you were away. Solar cooking is based on this movement of the sun. We're going to learn solar time from the movement of shadows. Look at the beans. Are there any signs of small bubbles near the top of the water?

Reset the cooker, again about an hour ahead of the sun and go about your chores. Now, you might want to watch it a little closer. Note how small bubbles near the top of the water progress over time to a much more active boiling. When the beans have been boiling for a while, take another jar and put a couple of cups of water in it. Set it next to

Chapter 8: Using the SunStar

the jar of beans. Refocus the cooker ahead of the sun again, by an hour or so.

Watch how the cooler water next to the beans slows down, but does not stop, the boiling of the beans. Some of the heat from the beans is now going to the other jar. This is one way to adjust the rate of boiling, keeping the food simmering.

After 45 minutes to an hour, open your cooker. Using a hot pad, open the lid on the jar of water. It should be steaming now, and ready for a cup of some kind of grain, like rice or millet. When you've added the grains, close the lid loosely. Set up the cooker again, and aim it ahead of the sun. Soon, the grains should be boiling also.

At this point, you may want to cut up some veggies to add to your meal. There are different ways you can handle vegetables. Onions and seasoning might be added to the beans. You can open the jar with the grains, when most of the moisture is absorbed, being careful of the hot steam. Fill it the rest of the way to the top with veggies. Sometimes, at this point, I like to add a little butter, margarine, or olive oil to the rice, then set the cooker back up and let it finish cooking. This method can steam veggies to perfection.

Another way to solar cook veggies, is to put them in another jar. You can let these veggies cook in their own juices, or add a small amount of water. Then put the lid on the jar and set this jar behind the other jars. Set your cooker back up and refocus it ahead of the sun. When it looks done, bring it out to eat, or close the reflectors over the glass to keep the meal hot until you're hungry.

Take time to savor and enjoy the first of many solar cooked meals, but also learn from the meal. How did it turn out? Were some things not cooked enough, or over cooked? Did the food need more or less water? Could you have cooked larger quantities, or should you have used smaller amounts? How would you change your cooking methods next time? Let the child experimenter in you come out. Follow your intuition and experience.

Heaven's Flame

Here's another important way to learn how to use your SunStar. Many foods can cook while you're away from home. Potatoes are an easy one to learn from. Put the potatoes in a jar and into the cooker. Point the cooker in the direction of the early afternoon sun. Now go away and come back after work or play. As the sun comes onto focus, the food will rise to the cooking temperature.

When you come home, the potatoes should be done. One of the variables you will learn to use while cooking this way will be the amount of food you cook. If the food is not quite done, next time use less food. This method of cooking should be used when you're pretty sure no afternoon showers will be developing while you're away. Try other foods using this method.

This leads to a third method of cooking, which will be a blend of the first two. In the morning, put some beans and water in your cooker and keep it focused until it approaches boiling. Then, if you have to leave, set the cooker well ahead of the sun. The sun will have one more pass to cook while you are away. This can be a very good way to cook long cooking beans.

To become very familiar with the potential of your own cooker, it's a good idea to find out how much food it can possibly cook in a day. Pick a day with good sun, a day when you will be around to keep it focused. Cook larger quantities than you have done before and see how they turn out. This will be very useful information to you in being able to effectively use your particular design. This maximum cooking potential will vary during different seasons.

These cooking methods, and variations of them, will allow the solar cooker to fit into your own lifestyle.

Learn about the many variables that affect cooking times. One of these, which I just mentioned, is the amount of food you put in your cooker. It will take twice as long to cook twice as much food and less time for smaller amounts. Also, different foods take different times. Another variable will be related to the cooker you built. A larger cooker can cook larger amounts of food in the same amount of time. The other variables will come from your own environment. The greatest variable will be the sun itself. As the Eskimos have many words for snow, the solar cook will develop an understanding of the many variables relating to the power of the sun. A bright, cold, dry day might have more cooking power available than a much hotter, humid day. You may be able to cook faster in a high mountain area than in a coastal area, again due to humidity. Also, there are many variations in the sun, according to the season of the year. The summer has longer cooking hours than the winter.

Watch the clouds. When do they build up? As you become better at solar cooking, you will have fun dancing around these clouds, cooking in the morning before they build up or later after they burn off. Note which seasons have consistently good sun in your area.

You may live in a very polluted area, so cook when the sun is high so that the sunlight does not have to travel through as much smog. The smog robs sunlight of its power. Sometime in the future, we will demand the right to clean air, so that our children can breathe freely, and so that our solar cookers can cook with full power. If you live in these smoggy areas, be sure you take your cooker with you when you travel to cleaner areas. See how the cooker works differently with less smaze. Realize what you are missing by allowing the air to be so dirty.

This is all a learning experience. Learn from failures, but let your successes guide you and lead you into the solar culture. Accept the limitations of solar cookers. Realize all things in nature have limitations. The goal of solar cooking is learning to work with the natural flow. Let it be a dance of joy and wonder.

Heaven's Flame

Chapter 9

Cooking Tips for Various Foods

This section will not be a like a regular cookbook with recipes. Instead, I will give you tips on cooking various foods. With this, and a little experience on your part, you will be able to use any recipe book. You will learn how to adapt recipes to your cooker with experience.

Solar cooks must take a more active role in cooking. There are many variables you will be dealing with, from the conditions of the day, to how often you can be around your cooker. The solar cook must gain an intuitive understanding of the art of solar cooking. Then, many recipes from different sources will be useful.

Solar Cooking Rules of Thumb

Start cooking early.

Try not to start cooking as the sun is lowering towards sunset.

More food = more time; less food = less time.

Different foods = different cooking times.

A large amount of quick cooking food takes the same time as a small amount of long cooking foods.

Sun high in the sky = more power; sun low in the sky = less power

Bright sun = more power; hazy sun = less power.

Summer solstice has the longest cooking hours; winter solstice has the shortest time for cooking.

Spring and fall equinoxes have the same cooking hours.

The quality of the sunlight is more important than outside temperature.

No stirring is necessary.

Use hot pads; food will be hot.

Use dark cooking pots.

Use lids on pots for faster cooking.

Recycled jars painted black make very good cooking vessels.

Don't overfill cooking jars; food expands.

Keep lids on jars, but don't tighten them down so that steam can escape.

Keep experimenting with what you can cook.

If you have a failure, try to understand why, change your technique and try again.

Learn to aim the cooker ahead in the path of the sun.

Prop the cooker with whatever is handy, in front and in back so that the wind won't tip the cooker over.

Left: Bread baking in coffee cans.

Right: Perfect can bread baked by the sun.

Bread

To bake a traditional bread loaf, set the pan in the ovenbox so that the heat will be able to circulate around it. Otherwise, the bread will cook on top but not the bottom. One trick that may help you cook traditional breads is to use a heat sink. Before you start to mix your dough, put a heat sink in your cooker. This could be a dark rock, or a piece of metal weighing a few pounds. Set the cooker up pointing towards the sun. As you work your dough, the heat sink will absorb heat from the sun. After the bread has risen, put the loaf in your cooker on top of the heat sink, which should now be hot. This way, the bread cooks from below, using the stored up heat, and from above, using the sun. With this method, I usually like to cook the bread when the sun is high, so I have maximum power.

Another way to cook bread is by making rolls. These are not as thick as a regular bread loaf, so the heat is able to easily penetrate the rolls. There is no problem with the bread not cooking on the bottom.

My favorite way to cook solar breads is in 1 or 2 pound coffee cans. Darken the outside of the can by painting it or burning it in a fire. If the cans are already a dark red or brown, they will work without any additional darkening. After you use these cans for a while the insides will naturally darken and cure like regular baking tins.

In the USA, it has been illegal for a number of years to use lead seals in tin cans, so you don't have to worry about lead contamination. This may not be the case in some countries, so be careful. As I remember, on the old tin cans that used a lead seam, you could see the solder easily. In the USA, they now electrically seal the joints. These new joints look very different than the old lead seals.

To bake the bread in tin cans, first mix up and knead the dough. If you don't have a warm place to raise your dough, you can put it in the ovenbox with the glass on, but without the reflectors. Aim the cooker so that just a little sunlight gets in the box. You don't want the dough to get above 120° F, or it will kill the yeast. At 120° F, it becomes too hot to comfortably touch with your hand. Keep it warm, but well below hot.

When the dough rises, punch it down and knead it again. Oil your coffee tins, and fill them about 1/2 way to the top with dough. Next, place two or three of these cans in your cooker on the rack, and point the solar cooker to where the sun will be in one hour. Leave the reflectors off, or back and to the side. When the dough starts to actively rise, put the reflectors on and let the sun pass over.

By the time the sun passes, the bread should be done, depending on the amount of sun and how much dough you put in. At this point, you can remove it, or point the cooker ahead of the sun again to give it a harder crust. Sometimes, I rub a little butter or margarine over the top of the bread to develop a softer golden crust. When the bread is done, it shrinks slightly. With a little tapping it should slip free of the ribs in the coffee can. If the bread is not done, it won't slip out. Put it back in the cooker for a while longer.

Please try using cans to cook bread. I have had my most regular successes with solar bread using tin cans. Something about the shape of the can and the thinness of the metal cooks the bread very well right down to the bottom of the loaf. It makes very nice moist bread that can be cut into thin slices for sandwiches. Another good way to eat it is to cut slices from your tin can loaves. Butter them and put them back in the solar cooker for a short time. Try it, it's excellent.

Chapter 9: Cooking Tips

Beans (or Legumes)

Beans are a staple of my family. We have found that there is no better way to cook beans. There are flavors that come out of solar cooked beans that you can not duplicate with any other way of cooking. You will have to trust me until you've solar cooked them yourself. Sometimes, it's hard for my family to eat beans cooked other ways now because they taste like cardboard compared to solar cooked beans. The flavors are just not comparable.

There are many different kinds of beans to cook, so they can become the base for a wide variety of meals. Because the cooking times of different beans vary so widely, there is room to adapt different beans to different conditions. In the morning, as you step outside to greet the day, a bright sunny morning might mean a pinto day. Put some beans in your cooker early, even if you don't know what you want to eat later. The beans will work as a heat sink, capturing the morning sun. As they come up to boil, you can build a meal around them. Even if you decide later that you don't want to eat the beans that day, you can put them in your refrigerator. Later, you can reheat them, maybe using traditional cooking methods to create burritos or other foods.

The ratio of water to beans is 1 cup of dry beans to 2 1/2 cups of water. This ratio may vary depending on whether you are simmering or actively boiling the beans. You will learn to adjust the water ratio after you work with your cooker for a while.

Try not to run out of water, because the beans on top will dry out. The dried beans will not cook right even if you add water later. If you have to add more water to your beans, be careful—the beans will be steaming hot. Also, when you pour cold water into a hot jar, you run the risk of breaking the jar. Try to notice ahead of time if it looks like you may need to add more water. You can put a jar of water in the cooker to heat up. Then you can add warm water to the boiling beans. It is best if you learn to put the right amount of water in the jar initially, so you don't have to add more later.

You will also have to take into account the various cooking times of different beans. Generally, the large beans, like pinto or kidney beans, take the longest times. Smaller legumes, like black beans, split peas, lentils, or black eyed peas, cook faster. There are exceptions to this size rule though. For example, small red beans are a long-cooking bean.

When people cook beans, they often soak them overnight to shorten the cooking time. This can be awkward with solar cooking. It is often hard to know if it will be sunny the next day. Over the years of cooking beans with my solar cooker, I have found that it is not really necessary to presoak the beans anyway. There is something about the way the solar cooker slowly raises the temperature, degree by degree, that makes the beans cook just as fast without soaking as when they are soaked. At least, I have never been able to notice any timing difference. Because it is easier not to presoak beans, I rarely do.

When the beans are cooked in the solar cooker, they end up very well-done and soft. I've fed solar cooked pinto beans to my toothless babies and they have had no problems eating them. The only reason you might want to soak beans is if you have problems with gas. Soaking is said to help this.

Once they are up to temperature, beans need to continue to cook. If you start cooking beans in the solar cooker, but clouds build up before they fully cook, it is best if you remove them and cook them with your traditional cooking tools. If beans almost cook, but are allowed to cool down, there is a potential that they will never cook right, no matter how long you cook them.

Grains

Grains are an important part of the diet and they cook well in a solar cooker. These include millet, barley, wheat, corn, and many varieties of rice. Also, some of the older grains like amaranth are being reintroduced into our diets, and can be cooked in your solar cooker.

Most of these grains can be cooked similar to regular cooking methods. Place a couple of cups of water (maybe slightly more) into a jar, and put it into the cooker to heat up. After about an hour, open the cooker. With a hot pad, open the jar and put in a cup of grain. Be careful of the hot steam. Set up the cooker again, pointing it ahead of the sun and cook until the grain is done. For fluffier grain, coat it with a tablespoon of oil before adding it to the hot water.

Long grain brown rice is one of the few grains that can be started in cold water and just placed in the solar cooker. This is useful if you want to cook rice while you are gone. Long grain rice uses 2 1/4 cups of water to 1 cup of grain.

With corn, polenta, or cornmeal, use a ratio of 5 to 1, water to corn. To cook the ground corn, pour 2 1/2 cups of water into a cooking jar, and place it in the solar cooker. In another bowl, not in the solar cooker, place 1/2 cup of corn. Stir in 1 cup of cold water. As the water in the solar cooker rises in temperature for the next hour or so, the corn is soaking up the cold water. When the water is hot, stir in the moistened corn, and cook until done. If the jar with the water is very hot, be careful when putting in the cold corn water mix. Try to pour it directly into the water, not down the side of the hot glass, or it could crack. If this becomes a problem, you might want to try heating up the corn water mix slightly before adding.

When grain does not fully cook on the bottom of the cooking jar, there are two possible reasons. One is that the grain needed more cooking time. The other more common reason, is that a little more water was needed. What happens is that the grains on top soak up all the available water as they cook. By the time the grains on the bottom rise to cooking temperature, there is no more available water. Next time, adjust your original water ratio so that there is a little more water.

Vegetables

There are many different ways to cook vegetables in your solar cooker. They can be steam baked in their own juices, boiled in soups, or added to casseroles.

To steam bake most vegetables, put them in one of your solar cooking jars without any additional water and put them in your cooker. When they have that freshly steamed color, they are done.

Broccoli cannot be cooked in this way. There is too much of a cooking variation between the flowers and the stem. If you were to put broccoli into a jar, the flowers would overcook and darken before the stem would cook. Cook broccoli on top of steaming grains. When the grains are almost done, open the jar carefully, and add the broccoli. Other vegetables can also be cooked this way. They steam quickly using this method.

Artichokes, with a little additional water placed in the bottom of a cooking jar, can also be solar steam baked to perfection. Obviously, you will need to use wide mouth 1/2 gallon jars to fit these in.

Tomatoes can be cooked in jars with other veggies and herbs, to make excellent tomato sauce.

Winter squash will steam bake in its own skin, if it can fit into the cooker without being cut it in to pieces. If you have to cut up the winter squash, you will have to cook the pieces in jars. One of the excellent winter squashes to solar bake, is spaghetti squash, which then can be substituted for noodles. Try it with your favorite spaghetti sauce.

With potatoes, you can also bake them by oiling the skins and placing them next to other boiling jars, or in the oven by themselves. If you just place oiled potatoes in the cooker, the thin skin of the potatoes will leak out steam that could have been used for cooking. This will slow up their cooking, and they won't cook as well. Potatoes can be put whole or cut up into jars for cooking. To tell when potatoes are done, look for a little build up of water on the bottom of the cooking jar. This makes excellent steam baked potatoes—the kind that would be good in potato salad, or saved for the morning's hash browns.

One interesting phenomenon of solar cooking can be seen when you cook different sized potatoes in the same jar. A large potato will cook at the same rate as a small one.

Explore new methods of cooking vegetables. This will greatly expand what you can do with your cooker.

Other Foods

Pizza: cook the crust first, then put on sauce, veggies, and cover with cheese. The pizza is done when the cheese starts to bubble.

Pies: cook the filling first, in a jar; then cook the bottom crust. Put the filling in the crust, and add a top crust, if you want, and bake again.

Cheese melts: good, quick snack food. Put a piece of bread with the same toppings as a pizza, or cheese alone, and cook.

Calzones: became one of the favorite foods of our family after we learned that they cooked so well in a solar cooker.

Nachos: chips, salsa, and cheese. Cook until cheese melts.

Omelets: can be cooked in a solar cooker. Hard boiled eggs can be cooked without water. The only problem is how to tell when they are done.

Cinnamon rolls: bake in a muffin or biscuit pan. Arrange so that the heat can circulate around the pan.

Cookies: same as any cookie recipe but they take a little longer because you don't preheat the cooker.

Cobbler: these make excellent taste treats in a solar cooker.

Cornbread or other quick breads: turns out very moist and good.

Cakes: fun for special birthdays.

Granola: cook the same as usual, but take it out as soon as it starts to darken; it's easy to overcook.

Noodles: boil water while noodles, dry in another jar, heat up. Add hot noodles to boiling water.

There are many types of food that I don't cook that will work well in a solar cooker. This is why I want to stress that, all of us are experimenters. Out of our collective efforts, we will see what these cookers can do. Take what I have said here as tips. Try them out, but develop your own understanding of how solar cookers work. Apply your own experience to your favorite recipes, to make the solar cooking experience uniquely your own.

Chapter 10

How to Produce the SunStar in Kit Form

I have always found it relatively easy to hunt for the right sized boxes to make my own personal cookers, but if I tried to build more than just a few, the search for the right size boxes became quite time consuming. This has led me to develop a workshop kit of the SunStar. The goal of this kit is to educate people on how to build the SunStar solar cooker. Included in this kit are only the important cardboard pieces to make the SunStar. It is a workshop kit because the person putting it together will be doing all the work except the search for the supplied boxes. By following a step by step plan to build this cooker, they will acquire all the necessary skills to build more cookers on their own. They will end up with a very useful sized cooker and their first attempts at cooking with it will more likely be favorable. Later, when they go on to build their own cooker from scratch, they will have something to compare it to. I want to share with you how to put this kit together, because it could be a very effective way to get this cooker out to many people.

In most medium to large cities, there are companies that sell cardboard boxes to the public. Look them up in the Yellow Pages under Boxes, or Corrugated Boxes. The companies will give you a pamphlet of sizes they have available. The minimum order for boxes from these sources is usually in packets of 25 at a time.

To build my favorite SunStar in kit form, order the two boxes that define the ovenbox as well as four flat pieces to make the reflectors. The sizes of these boxes are 12" x 12" x 10", (L x W x D) and 14" x 14" x 12". The larger box is

Above: Caribbean school children with their "SunStar."

easy to find in health food stores because it is used to hold four 1 gallon bottles of organic apple juice. The smaller box is harder to find. This is the reason that I developed this kit.

I chose these particular boxes because a traditional baking pan fits perfectly for a rack. This pan is not included in the kit, but many people will find that they already own one. If not, it can be found used at thrift stores. It is sold by Ekko and Baker's Secret. The size is listed as 10 3/4" x 7" x 1 1/2". With the handholds, its actual length is just over 12". By bending down the handholds slightly the pan slips

Chapter 10: The SunStar Kit

snugly into the 12" x 12" inner ovenbox. This is a common size, but you may have to check a few stores. It comes in a dark gray color that will naturally darken with use.

For the reflectors, the cardboard companies sell what they call pads, which are flat pieces of cardboard. The pad size you will need is 24" x 30" (L x W). Unfortunately, they don't sell this size, so you will need to buy the 24" x 36" pad and cut off the extra six inches.

The prices will vary slightly. At the time of writing this, the two boxes in packets of 25 and the 100 flat pieces of cardboard necessary to build 25 cookers comes to about five dollars per oven.

To create the reflectors, stencil solid cutting lines and dotted bending lines on one of the flat pieces of cardboard. This piece is then used as a pattern to make the remaining three reflectors.

The kit does not include the glass for a few reasons. The glass would add to the cost of the kit, make it heavier, and might break before ever being used. Many people have never bought glass and are intimidated by the prospect. It is important that they know that glass is easy to buy. Otherwise, some people would think that their solar cooker was no good anymore if the glass broke. In the kit, I would tell them it is very easy to go to a glass store to buy their own glass. For this kit, the glass would be 12.5" x 12.5".

The oven kit uses recycled cardboard boxes to be cut up for the insulation. This way, the kit still ends up being built by about 50% recycled boxes. The sizes of these boxes are not important, so they are easy to find.

I'm sharing this with you because at this point we need to start getting cookers out in large quantities. With your help, the SunStar kit can be a way to do it. I hope that many of you might try your hand at producing these kits, maybe even picking up a little extra cash for your effort. Maybe you are a part of a group of people who want to try this cooker out, and making a collective buy of cardboard would make it worthwhile. With these kits, many different types of community groups could sponsor cooker building workshops to stimulate interest in solar cooking.

If you are interested in this idea, I'd suggest that you build a prototype of this cooker using the plans in this book. The outer box, 14" x 14" x 12", is easy to find at health food stores. The reflectors will also be easy to find from large appliance boxes. The only box that may be hard to find will be the inner box. With a little searching you should be able to find one that will work. It seems that there are common boxes of the size 12 1/2" x 11 1/2" x 10", and this box could be used. Although this box is not quite square, for the purpose of building reflectors, just assume that it is.

With this chapter, I reach the end of what has been a long adventure. I hope the effort was worth it, and that it gives you the knowledge and inspiration to make a solar cooker. Solar cookers could have a profound effect on our culture, giving all of us a simple way to exercise our ability to change in a positive direction.

At some point, there will be many people using solar cookers. The knowledge and ability to solar cook will become available to all people, and to all future generations, to use if they so choose. I believe that we will reach this threshold soon.

Chapter 11

The Rainbow Cooker

While finishing this book, I built a larger cooker that I'd been thinking about for a while. It worked out so well that I need to share it with you before this book is published. This cooker does not really break any new ground from what has already been presented, but it does illustrate how you can build a specific cooker to fill a specific need.

I have been sharing my SunStar solar cooker at the National Rainbow Gatherings over the last few years. The SunStar, is a fairly small family-size cooker and was not able cook a large enough quantity of food to share. Instead, I told people to build their own, learn to use it, and bring it to the gathering the following year.

Knowing how much people are lead by their stomachs, I started to wish that I had a larger design that could feed more people. This large-sized cooker needed to fill certain design goals to function at a gathering. We usually hike a couple of miles to a gathering site, so it needed to be easy to transport. It also needed to be a fairly powerful oven to impress the folks with the amount and quality of food it could cook. And, like the SunStar, it needed to be built out of inexpensive, recyclable materials. I also wanted it to be very easy to use by one person.

The cooker I came up with succeeded in all of these goals. I was able to cook 250 bread rolls a day. It also produced pizzas, cornbread, cakes, cookies and corn on the cob during my stay at the gathering. The cooker can be set up or taken apart in 5 minutes using only 4 wingnuts and a few shoestrings. When disassembled, this cooker folds com-

Heaven's Flame

pletely flat for easy transport and storage. The cooker was built using basic materials: cardboard, glass, foil, a little wood, and some silicone. The total cost to put this one together was twelve dollars. Six of those went to the dump and six were spent on foil. Some of the materials, like the white paper glue, I already had on hand.

This cooker's design started with the recycled glass which I found at the dump. This was important because the glass, if bought new, would have been very expensive. At the dump, I found a screen door with two pieces of safety glass of the same size. This allowed me to build a thermal pane, with some wood strips and silicone. Safety glass is much stronger than regular glass, so the likelihood of breaking it is lessened. I left a small gap in the thermal pane seal to let any trapped moisture escape.

Below: The "Rainbow" cooker (left) next to a "SunStar."

Chapter 11: The Rainbow Cooker

*Left:
The "Rainbow" opened for access to food.*

*Below:
Collapsed for portability.*

Next, I focused on the ovenbox. As you see from the pictures, I based it on a triangle shape which is very stable and strong. The two side triangle wedges are able to separate using the four wingnuts, which allows the ovenbox to fold flat for packing away. The oven walls used some wood and one inch of layered cardboard and foil for its insulation.

The rack is a cut down oven rack with a chain, as described earlier in the book. The stand uses some 2x4's, rope, and a bolt pivot point. Both are fairly primitive methods, but they work and that is the goal.

The reflectors use two refrigerator boxes and are built just like a larger version of the SunStar's square cardboard reflectors. They are 3 feet in depth, and use two strips of 18" wide aluminum foil. When applying the foil to these

*Left:
The "Rainbow"
in cooking
mode.*

larger surfaces, I found it easier to wet the cardboard and the dull side of the foil with the glue/water mix. Then I applied a flat weight to the reflectors so they stayed flat as they dried. The weak points of the reflectors were strengthened with glued cotton cloth and sewed with shoe strings.

The glass used for this cooker was not square, but slightly rectangular, at 27" x 32". I wanted to use the square version of my reflector style because I really wanted them to fold up together without having to separate two corners. To accomplish this, I lengthened the side triangular pieces until they were 32" wide and based the reflectors on a 32" square.

I don't want to get into a more detailed description of this cooker because it is just a prototype. The next time I build one, it will probably be built slightly different. I just needed to share it with you before this book was published because it shows adapting designs to meet different needs.

At the gathering, I arrived about a month early when there were just a few people. When I came on site and set up I was the first bakery. At this point, I significantly added to the food supply and everyone was talking about the solar

Chapter 11: The Rainbow Cooker

bread. As the numbers of people increased, this cooker became more of a neighborhood cooker.

During the gathering, our numbers swelled to over twenty thousand hippies in the woods. Wood fires are traditionally the main cooking source when dealing with high numbers of people that need to eat. I started to perceive that there could be a place for solar when dealing with this many people. Wood fires can bring food quickly up to cooking temperature, but holding the food at a simmer is difficult, causing smoky fires and kitchens, and often scorched foods. While solar takes a longer time to bring food up to boiling, it is perfect for keeping food at a simmering temperature, with no smoke or scorching. Using a combination of wood and solar could work well.

I envision using an efficient wood-burning rocket stove to bring a series of 4 or 5 large pots of food up to a boil early in the morning. The rocket stove uses an 8" thin walled pipe which is rocked and mudded into a easy to use stove. Small sticks can then be burned with high efficiency and no smoke. When it is burning correctly, it makes a slight roaring noise, hence the name rocket stove.

When each pot reaches boiling from the wood fire, it can be added to a large slant faced solar cooker to keep it simmering throughout the day. The solar cooker could use a thermal pane sliding glass door as its glazing. This glass would be built into an insulated box at a 30 degree angle. The oven box could have a series of 4 or 5 doors in back to put the boiling pots inside. This cooker could have two simple adjustable reflectors connected to the long side of the window of the ovenbox. For ease, it would be nice to have wheels on this cooker, set up like a little red wagon. I believe this could put out a large quantity of high quality food. If clouds build up, you could close down the reflectors and insulate the glass so that the food would continue to cook in the insulated box as it slowly lost temperature.

Heaven's Flame

Access

Most of these groups and individuals will enjoy hearing from you. If you are looking for a returned letter with information, it would be nice if you sent them a self-addressed stamped envelope (SASE). Also, you might consider sending a small donation to cover any printing costs or office work. I know that most of these people are not well funded at this time, and their resources are probably tight with the work of educating people about solar cooking.

A SASE will be nice if you contact me, and any tip for my effort would also be appreciated. I've always enjoyed answering my mail, but sometimes postage and printing costs push my budget beyond its limit.

Heaven's Flame
Joe Radabaugh
PO Box 111
Mt. Shasta, CA 96067

Avinashilingam Institute for Home Science and
Higher Education for Women (Deemed University)
Coimbatore 641 043
Tamil Nadu, India
E-mail: devunity@giasmd01.vsnl.net.in
Web: www.avinashiligam.edu

Roger Bernard
A.L.E.D.E.S., Universite
Lyon I, 69622 Villeurbanne, France

Citizens For Solar, Tucson Solar Potluck
PO Box 36744-6744
Tucson, AZ 85740
Phone: 520-292-9020
E-mail: billcuningham@dakotacom.net

Bud Clevett's Sunspot Solar Oven
Distributed by Hubbard Scientific
PO Box 2121
Fort Collins, CO 80522
Phone: 800-289-9299

Cooking with the Sun
Morning Sun Press
PO Box 413
Lafayette, CA 94549
E-mail: jdhowell@ix.netcom.com
Web: http://home.ix.netcom.com/~jdhowell

EG Solar e. V.
Neuoettinger Str. 64 c
D-84503 Altoetting, Germany
Phone: 0049-8671-96 99 37
Fax: 0049-8671-96 99 38
E-mail: eg-solar@t-online.de
Web: www.eg-solar.de

Gruppe ULOG
Morgartenring 18
CH 4054 Basel, ULOG
Switzerland
Phone: +41 61 3016622

Barbara Kerr
The Kerr-Cole Sustainable Living Center
PO Box 576
Taylor, AZ 85939
Phone: 928-536-2269
E-mail: bkerr@cybertrails.com

Access

Bill Lankford
Central American Solar Energy Project (CASEP)
10718 Scotts Drive
Fairfax, VA 22030
E-mail: casep@erols.com
Web: http://mason.gmu.edu/~wlankfor/

Ed Pejack
School of Engineering, Baun Hall
University of the Pacific
3601 Pacific Avenue
Stockton, CA 95211
E-mail: epejack@yahoo.com

Solar Cookers International
1919 21st Street, #101
Sacramento, CA 95814
Phone: 916-455-4499
Fax: 916-455-4498
E-mail: info@solarcookers.org
web: www.solarcooking.org

Solar Energy International
PO Box 715
Carbondale, CO 81623
phone: 970-963-8855
Fax: 970-963-8866
E-mail: sei@solarenergy.org
Web: www.solarenergy.org

Solar Freedom International
402 24th Street West
Saskatoon, Saskatchewan S7L-0B8
Phone: 306-652-1442
Fax: 306-652-1442

Sun Ovens International Inc.
Manufacturer of Tom Burns' Sun Oven
39W835 Midan Drive
Elburn, IL 60119 USA
Phone: 630-208-7273
Toll Free: 800-408-7919
E-mail: sunovens@execpc.com
Web: www.sunoven.com

Richard C. Wareham, Director
Sunstove Organization,
3140 N. Lilly Road,
Brookfield, WI 53005
Phone: 262-781-1689
Fax: 262-781-0455
E-mail: sungravity@attglobal.net

Sunstove
PO Box 21960
1515 Crystal Park
Republic of South Africa
Phone: +011 969 2818
Fax: +011 969 5110
E-mail: sunstove@iafrica.com
Web: sungravity.com

Zomeworks Corporation
1011 Sawmill Rd. NW
PO Box 25805
Albuquerque, NM 87125
Phone: 800-279-6342
Web: www.zomeworks.com

Liu Zude, Research Professor
Henan Energy Research Institute
29 Huaguan Road Zheng Zhou
Postcode 450003
Peoples Republic of China

I can't guarantee that this information will stay current. There are probably other great resources that have been overlooked, or are new since publication. Check out www.solarcooking.org for up to date info on manufacturers of solar cookers and other books on solar cooking.

I have non-copyrighted, one page plans for the SunStar solar cooker. They are great for sharing with friends. I'd like to consult with anyone producing SunStar Kits, so that I can be sure you have latest information and building techniques. Also, if anyone wants to publish this book in a foreign language, please contact me. We are also looking for donations towards a traveling fund so that my family and I can go on a farmers market, solar cooking educational tour, hopefully worldwide.

I am now beginning to collect information for the next edition of this book. Keep me informed on your solar designs or promotions. Thanks.

Heaven's Flame

Index

Addresses 137
Avinashilingam Institute for Women 48
Baer, Steve 63
Beans 121
Bernard, Roger 37
Bluhm, Bev 43
Boxes, variables 88
Boxes, altering 108
Bread 119
Burns, Tom 55
Canning 32
Central American Solar Energy Project (CASEP) 49
Citizens for Solar 69
Clevett, Bud 53
Climate 26
Clouds 28
Cole, Sherry 42
Convection Cooker, type 67
Cooking temperatures 24
Cooking times 24
Cooking with the Sun, book 54
Cookit, type 38
Cookware 30, 106
Eaton, Ed 48, 69
Eccentric Axis, type 64
EG Solar 61
Erwin, Sam 59
Focusing the cooker 29, 104
Freedom Cookers 58
Froese, Joe 57
Glass, double pane 79
Glass, variables 78
Glass, why it cracks 12
Glass, why it steams 13

Glazing, variables 78
Grains 122
Grupp, Michael 68
Gruppe ULOG 44, 65
Gurley, Heather 51
Halacy, Dan 54
Heat Sinks 16
Hybrid Kitchen, type 65
Insulation, variables 76
Jars as cookware 30, 106
Kerr, Barbara 37, 42
Lankford, Bill 49
Larson, Bob 51
Lippold, Christine 65
Materials, for SunStar 90
Metcalf, Bob 43
Olehler, Lisa and Ulrich 44
Ovenbox, variables 75
Oven door, variables 76
Paint, variables 81
Parabolic, type 61
Pizza 125
Plastic glazing 78
Pots 30, 106
Preheating 26
Rack, variables 82
Rainbow Cooker, type 131
Reflectors, variables 83
Salad Bowl Cooker, type 36
Scheffler, Wolfgang 65
SERVE 46
SK12, type 61
Solar Box Cooker, type 41
Solar Chef, type 59
Solar Cookers International (SCI) 37, 42
Solar Cooking Naturally, book 52

Index

Solar Energy International 47
Solar Panel Cooker, type 36
Solar Science Projects, book 54
Sun Oven, type 55
Sunflash, type 63
Sunlight Works 51
Sunspot, type 54
SunStar, type 87
SunStar, as a kit 127
Sunstove, type 39
Stand, variables 82
Stirring 27
Stone, Laurie 48
Taste, of solar cooked food 23
Telkes, Maria 15
Temperatures 24
Thermal pane glass 79
Tucson Solar Potluck 69
United Nations, studies 2
United Nations, Summit on the Environment 6
Vegetables 123
Villager, type 56
Wang, Xiping 64
Wareham, Richard 39
Water purifying 9, 44
Water ratios 27
Wind 104
World Conference on Solar Cooking 44
Yoder, Lui 63
Zomeworks 63
Zude, Lui 64

Colophon

This book is printed on recycled paper with soy based ink. The paper, Environment® PC 100, produced by Neenah Paper, is made of 100% post-consumer paper fibers. The interior pages are 80 pound text in natural. The cover is 80 pound cover in white.

Home Power Publishing supports the use of alternative fibers in the production of paper. Fibers such as hemp, knaff, and bamboo are fast-growing, environmentally sustainable alternatives to tree fiber—and they make great paper! These fibers can and should be produced in the locally. Until they are more readily available, Home Power Publishing commits to using high post-consumer content, recycled papers.